An Introduction to the Theory of Groups

Paul Alexandroff

Translated by
Hazel Perfect
and
G. M. Petersen

Dover Publications, Inc.
Mineola, New York

Bibliographical Note

This Dover edition, first published in 2012, is an unabridged republication of the work originally published in 1959 by Blackie and Son, Ltd., London and Glasgow. That English translation was made from the German *Einführung in die Gruppentheorie*, published in 1954 by Deutscher Verlag der Wissenschaften, Berlin, which is itself a translation from the Russian of Paul Alexandroff by Lothar Uhlig.

Library of Congress Cataloging-in-Publication Data

Aleksandrov, P. S. (Pavel Sergeevich), 1896–1982.
[Vvedenie v teoriiu grupp. English]
An introduction to the theory of groups / Paul Alexandroff ; translated by Hazel Perfect and G.M. Petersen. — Dover ed.
p. cm.
Originally published: London : Blackie, 1959.
Includes bibliographical references and index.
ISBN-13: 978-0-486-48813-4 (pbk.)
ISBN-10: 0-486-48813-6 (pbk.)
1. Group theory. 2. Catalysis. I. Title.

QA174.2.A43513 2012
512'.2—dc23

2011045172

Manufactured in the United States by Courier Corporation
48813602
www.doverpublications.com

Translators' Foreword

The group concept has played a central part in the development of mathematical thought. At the same time, many aspects of the theory of groups are simple enough for the interested grammar school student to appreciate and absorb. We believe that this book by the eminent mathematician P. S. Alexandroff is especially suited for introducing the subject to the young student. The material presented is of fundamental importance and is developed in a clear and rigorous fashion, and the book is particularly noteworthy on account of the wealth of illustrative examples which are included.

We have added some exercises at the end of each chapter, and some references which we hope will be useful to the English-speaking student. We have also made a few minor alterations and additions to the actual text, some of which were made necessary by English usage, and we have corrected some minor errors.

Throughout the work we were able to consult Mrs. A. M. H. Marleyn over language obscurities, and we owe her a great debt of gratitude for the valuable and generous help which she gave us. We are also very grateful to Dr. H. O. Foulkes of the University College of Swansea for his critical reading of the text at the proof stage.

H. P.

G. M. P.

from the Foreword to the First Edition

Next to the concept of a function, which is a most important concept pervading the whole of mathematics, the concept of a group is of the greatest significance in the various branches of mathematics and in its applications. The group concept is not any more difficult to appreciate than the function concept; indeed one can more easily become familiar with this concept during the early stages of a mathematical education than with the subject-matter of elementary mathematics.

Every pupil in a senior class of a grammar school who enjoys doing mathematics is capable of grasping the idea of a group if he is interested and industrious. And so this book has been written in the first place for the mathematically inclined pupils in the senior classes in the grammar school, but also for those who teach mathematics to the senior or to the advanced level. As regards the character of the exposition, I have been at pains to introduce no concepts without illustrating them by means of simple examples, for the most part geometrical.

P. S. A.

Contents

CHAPTER FOUR. CYCLIC SUBGROUPS OF A GIVEN GROUP

CHAPTER FIVE. SIMPLE GROUPS OF MOVEMENTS

CHAPTER SIX. INVARIANT SUBGROUPS

CHAPTER SEVEN. HOMOMORPHIC MAPPINGS

CHAPTER EIGHT. PARTITIONING OF A GROUP RELATIVE TO A GIVEN SUBGROUP. DIFFERENCE MODULES

APPENDIX. ELEMENTARY CONCEPTS FROM THE THEORY OF SETS

Chapter I

THE GROUP CONCEPT

§ 1. Introductory examples

1. Operations with whole numbers

The addition of whole numbers * satisfies the following conditions, which we call *axioms of addition* and which are of very great importance for all that follows:

I. *Two numbers can be added together* (i.e. to any two arbitrary numbers a and b there corresponds a uniquely determined number, which we call their sum: $a + b$).

II. *The Associative Law*:

For any three arbitrary numbers a, b, c *we have the following identity*

$$(a + b) + c = a + (b + c)$$

III. *Among the numbers there is a uniquely determined number* 0, *the zero, which is such that for every number* a *the relation*

$$a + 0 = a$$

is satisfied.

IV. *To every number* a *there corresponds a so-called inverse* (*or negative*) *number* $-a$, *which has the property that the sum* $a + (-a)$ *is equal to zero*:

$$a + (-a) = 0$$

Finally yet another important condition is satisfied.

V. *The Commutative Law*:

$$a + b = b + a$$

* By the whole numbers we understand all positive and all negative whole numbers together with the number zero.

2. The rotations of an equilateral triangle

We show that it is possible to add not only numbers but also many other kinds of things, and that the above conditions remain satisfied.

First Example.—We consider all possible rotations of an *equilateral triangle* ABC about its centroid O (fig. 1). We agree to call two rotations identical if they only differ from one another by a whole number of complete revolutions (and therefore by an integral multiple of

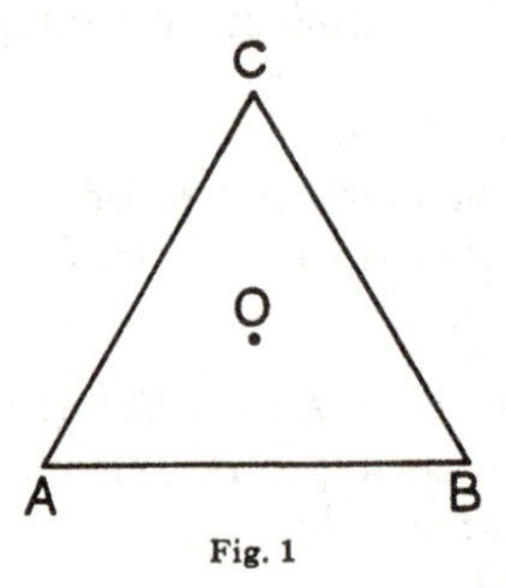

Fig. 1

360°*). We see without difficulty that of all possible rotations of the triangle only three rotations send it into coincidence with itself, namely, the rotations through 120°, 240°, and the so-called zero rotation, which leaves all the vertices unchanged and hence also all the sides of the triangle. The first rotation sends the vertex A into the vertex B, the vertex B into the vertex C, the vertex C into the vertex A (we say that it permutes cyclically the vertices A,B,C). The second rotation sends A into C, B into A, C into B, and therefore permutes A,C,B cyclically.

Now we introduce the following natural definition: The addition of two rotations means their successive application, the first rotation followed by the second. If we add the rotation through 120° to itself, then the result is the rotation through 240°; if we add to it the rotation through 240°, then the result is the rotation through 360°, the zero rotation. Two rotations through 240° result in the rotation through $480° = 360° + 120°$; their sum is therefore the rotation through 120°. If we denote the zero rotation by a_0, the rotation

* Since a rotation through an integral multiple of 360° sends every vertex back to its original position, we regard this rotation as identical with the zero rotation; more generally we interpret two rotations as identical if they differ from one another by a whole number of complete revolutions.

through 120° by a_1, the rotation through 240° by a_2, then we obtain the following relations:

$$a_0 + a_0 = a_0$$
$$a_0 + a_1 = a_1 + a_0 = a_1$$
$$a_0 + a_2 = a_2 + a_0 = a_2$$
$$a_1 + a_1 = a_2$$
$$a_1 + a_2 = a_2 + a_1 = a_0$$
$$a_2 + a_2 = a_1$$

Thus the sum of any two of the rotations a_0, a_1, a_2 is defined and is again one of the rotations a_0, a_1, a_2. We easily convince ourselves that this addition satisfies the associative law and evidently also the commutative law. Further, there exists among these rotations a_0, a_1, a_2 a zero rotation a_0 which satisfies the condition

$$a + a_0 = a_0 + a = a$$

for every rotation a.

Finally each of the three rotations has an inverse, which when added to the original rotation produces the zero rotation. The zero rotation is evidently inverse to itself: $-a_0 = a_0$, since $a_0 + a_0 = a_0$; further $-a_1 = a_2$ and $-a_2 = a_1$ (since $a_1 + a_2 = a_0$). Therefore addition of those rotations of an equilateral triangle, bringing the triangle into coincidence with itself, satisfies all the axioms of addition listed above.

We write out the law of addition of the rotations once more, this time in the convenient form of a table—an *addition table*:

	a_0	a_1	a_2
a_0	a_0	a_1	a_2
a_1	a_1	a_2	a_0
a_2	a_2	a_0	a_1

(I)

In this table we find the sum of two elements at the point of intersection of the row corresponding to the first element with the column corresponding to the second element.

If we wish to combine these rotations mechanically, then we simply take the three letters a_0, a_1, a_2 and add them according to the above table; moreover we can completely ignore the interpretation of the letters as rotations.

3. Klein's four-group

Second Example.—We consider the set of four letters a_0, a_1, a_2, a_3, whose addition is defined by the following table:

	a_0	a_1	a_2	a_3
a_0	a_0	a_1	a_2	a_3
a_1	a_1	a_0	a_3	a_2
a_2	a_2	a_3	a_0	a_1
a_3	a_3	a_2	a_1	a_0

(II)

or at length:

$$\begin{aligned} a_0 + a_0 &= a_0 \\ a_0 + a_1 &= a_1 + a_0 = a_1 \\ a_0 + a_2 &= a_2 + a_0 = a_2 \\ a_0 + a_3 &= a_3 + a_0 = a_3 \end{aligned}$$

$$\begin{aligned} &a_1 + a_1 = a_0 && a_2 + a_2 = a_0 \\ &a_1 + a_2 = a_2 + a_1 = a_3 && a_2 + a_3 = a_3 + a_2 = a_1 \\ &a_1 + a_3 = a_3 + a_1 = a_2 && a_3 + a_3 = a_0 \end{aligned}$$

Addition is defined for any two arbitrary letters of the set. We prove at once that this addition satisfies the associative and commutative laws.

The letter a_0 possesses the characteristic property of the zero that the sum of two elements, one of which is a_0, is equal to the other element.

It is therefore evident that the conditions I, II, III, V are satisfied in this " algebra of four letters ". In order to convince ourselves that condition IV is also satisfied, it is sufficient to refer to the relations

$$a_0 + a_0 = a_0, \; a_1 + a_1 = a_0, \; a_2 + a_2 = a_0, \; a_3 + a_3 = a_0$$

according to which each letter is inverse to itself (i.e. when added to itself produces the zero).

This " algebra of four letters " could appear at first sight as a mathematical game, a pastime without significant content. In reality the laws of this algebra expressed in Table II have a very real significance, with which we briefly acquaint ourselves. We mention moreover that this " algebra of four letters " is of great importance in higher algebra. It is called Klein's four-group.*

4. The rotations of a square

Third Example.—By means of considerations similar to those in the first example we can construct another " algebra of four letters " different from the one above. We consider a square ABCD and the rotations about its centroid which bring the figure into coincidence with itself. Again we identify any two rotations which differ from each other by an integral multiple of 360°. We have therefore altogether four rotations, namely the zero rotation, the rotations through 90°, through 180°, and through 270°. These rotations in this order we denote by the letters a_0, a_1, a_2, a_3. If we again understand by the addition of two rotations their successive application, then we obtain the following addition table, just as in the first example:

	a_0	a_1	a_2	a_3
a_0	a_0	a_1	a_2	a_3
a_1	a_1	a_2	a_3	a_0
a_2	a_2	a_3	a_0	a_1
a_3	a_3	a_0	a_1	a_2

In the same way as in the first example, we can consider rotations of a regular pentagon, or hexagon, or in general n-gon. It is left to the reader to carry through the appropriate details in this direction and to construct the corresponding addition tables.

§ 2. Definition of a group

Before we continue to consider other individual examples, we collect the results of the examples already investigated and introduce the following basic definition.

* After the great German mathematician Felix Klein (1849–1925).

We suppose given a certain finite or infinite set* G; further we assume that any two elements a and b of the set G define a third element of this set, which we call the *sum* of the elements a and b and which we denote by $a + b$. Finally we assume that this *operation of addition* (the operation whereby we proceed from two given elements a and b to the element $a + b$) satisfies the following conditions:

I. *The Associative Law. For any three elements a, b, c of the set G we have the following relation*

$$(a + b) + c = a + (b + c)$$

This means that if we denote by d the element of the set G which is the sum of the elements a and b, and similarly by e the element $b + c$ of the set G, then $d + c$ and $a + e$ are one and the same element of the set G.

II. *The condition for the existence of a null element. Among the elements of the set G there is an element which we call the null element and denote by* 0, *which is such that, for an arbitrary choice of the element a, we have*

$$a + 0 = 0 + a = a$$

III. *The condition for the existence of an inverse of each given element. Corresponding to any given element a of the set G we can find an element $-a$ such that*

$$a + (-a) = (-a) + a = 0$$

A set G with an operation of addition defined in it, which satisfies the three conditions listed above, is called a *group*. These conditions themselves are called *group axioms*.

If, as well as the three group axioms, the following condition is also satisfied in a group G, viz.

IV. *The Commutative Law*:

$$a + b = b + a$$

then the group is called *commutative* or *Abelian*.†

A group is called *finite* if it consists of a finite number of elements; otherwise it is called *infinite*. The number of elements of a finite group is called its *order*.

* See the appendix at the end of this book. In what follows we shall assume that the reader in quite conversant with the contents of this appendix.

† After the gifted Norwegian mathematician N. H. Abel (1802–1829).

Now that we have made ourselves familiar with the definition of a group, we see that the examples given in the first paragraph of this chapter are examples of groups. We have therefore so far become acquainted with the following groups:

1. The group of whole numbers.
2. The group of rotations of an equilateral triangle (this group is also called a cyclic group of order 3).
3. Klein's four-group.
4. The group of rotations of a square (cyclic group of order 4).

At the end of § 1 the rotation group of a regular n-gon was mentioned (cyclic group of order n). All these groups are commutative, and they are all finite with the exception of the group of whole numbers which is evidently infinite.

§ 3. Simple theorems about groups *

1. The addition of any finite number of group elements. The first rule for the removal of brackets

The associative law is of very great importance in group theory and also throughout algebra. By its means we can define the sum not just of two group elements but the sum of three elements, and in general the sum of an arbitrary finite number of elements; and in order to calculate these sums we can apply the usual rules for the removal of brackets.†

Let us suppose by way of example that three elements a, b, c are given, then *for the moment* we do not know what is meant by the sum of these three elements; for the group axioms speak only of the sum of two elements, and expressions of the form $a + b + c$ are not yet defined. But now the associative law states that if on the one hand we add the two elements

$$a \text{ and } b + c$$

and on the other hand the elements

$$a + b \text{ and } c$$

we then obtain one and the same element as their sum. Thus this element,

* If the reader wishes first of all to make himself familiar with other examples of groups, he may skip this paragraph and come back to it after reading Chapters II–IV.

† However we must realize that in the case of a non-commutative group the order of the elements which are added must not be altered.

which is the sum of a and $b + c$ and also of $a + b$ and c, may be *defined* without ambiguity as the sum of the elements a, b, c (in this order), and hence will be denoted simply by $a + b + c$. Thus we can regard the equations

$$a + b + c = a + (b + c) = (a + b) + c$$

as defining the sum $a + b + c$ of the three elements a, b, c.

We can conveniently define the sum of four elements a, b, c, d to be, for example, equal to $a + (b + c + d)$; and in this connection we prove that

$$a + (b + c + d) = (a + b) + (c + d) = (a + b + c) + d$$

First of all, from what has been stated above, we have

$$a + (b + c + d) = a + [b + (c + d)]$$

But for the three elements a, b, $c + d$ we have

$$a + [b + (c + d)] = (a + b) + (c + d)$$

On the other hand, for the three elements $a + b$, c, d

$$(a + b) + (c + d) = [(a + b) + c] + d = (a + b + c) + d$$

and this is what we set out to prove.

Now we assume that the sum of any $(n - 1)$ elements has already been defined; then we define the sum of the n elements $a_1 + \ldots + a_n$ to be $a_1 + (a_2 + \ldots + a_n)$, and we can therefore regard the expression $a_1 + \ldots + a_n$ as being defined for arbitrary n by the method of complete induction.

Theorem.—Let n be any natural number. For every natural number* $m \leqslant n$ *we have*

$$(a_1 + \ldots + a_m) + (a_{m+1} + \ldots + a_n) = a_1 + \ldots + a_n \quad (1)$$

Proof. The proof will proceed by the method of complete induction.† For $n = 1$ the theorem simply states that $a_1 = a_1$. We assume that it is true for $n \leqslant k - 1$, and prove it for $n = k$. We consider first the case $m = 1$. Then (1) becomes

$$a_1 + (a_2 + \ldots + a_k) = a_1 + \ldots + a_k$$

* A natural number is a positive whole number.

† It is recommended that the reader should first write out a proof himself, and then compare it with the one given in the text.

But this is just the definition of the expression $a_1 + \ldots + a_k$. Therefore formula (1) is true for $n = k$ and $m = 1$.

Now let us still consider $n = k$ and let us assume that our formula is proved for $m = q - 1$; we prove it for $m = q$. Since the formula (1) is obviously true for $m = n$, we can assume that $q < k$. Since the truth of the theorem for $n \leqslant k - 1$ is assumed, we have

$$(a_1 + \ldots + a_q) + (a_{q+1} + \ldots + a_k)$$
$$= [(a_1 + \ldots + a_{q-1}) + a_q] + (a_{q+1} + \ldots + a_k)$$

The associative law, applied to the three elements $(a_1 + \ldots + a_{q-1})$, a_q, $(a_{q+1} + \ldots + a_k)$, gives

$$[(a_1 + \ldots + a_{q-1}) + a_q] + (a_{q+1} + \ldots + a_k)$$
$$= (a_1 + \ldots + a_{q-1}) + [a_q + (a_{q+1} + \ldots + a_k)]$$

But the expression in square brackets on the right-hand side is by definition equal to

$$a_q + a_{q+1} + \ldots + a_k$$

Therefore we have

$$(a_1 + \ldots + a_q) + (a_{q+1} + \ldots + a_k)$$
$$= (a_1 + \ldots + a_{q-1}) + (a_q + \ldots + a_k)$$

But since the formula (1) is assumed to hold for $n = k$ and $m = q - 1$, the right-hand side of this last equation is equal to $a_1 + \ldots + a_k$. Therefore

$$(a_1 + \ldots + a_q) + (a_{q+1} + \ldots + a_k) = a_1 + \ldots + a_k$$

which is what we set out to prove.

2. The null element

The condition for the existence of a null element reads: *In the group there exists a certain element* 0 *such that for every element a of the group the condition*

$$a + 0 = 0 + a = a \tag{1}$$

is satisfied.

Nowhere does this condition contain the assumption that there can be no other element $0'$ different from 0 with the same property

$$a + 0' = 0' + a = a \tag{1'}$$

for every a. From the following rather more general proposition it results that in fact no such element $0'$ exists. We speak of this as *the theorem concerning the uniqueness of the null element.*

Theorem.—If corresponding to any element a of a group G there is an element 0_a, *which satisfies one of the conditions*

$$a + 0_a = a \quad \text{or} \quad 0_a + a = a$$

then it is necessary that $0_a = 0$.

Proof.—We assume first that $a + 0_a = a$, then it follows that

$$(-a) + a + 0_a = (-a) + a = 0$$

i.e.

$$0 + 0_a = 0$$

But from the definition of 0 we have

$$0 + 0_a = 0_a$$

Whence

$$0_a = 0$$

We can equally well deduce the identity $0_a = 0$ from the assumption $0_a + a = a$.

3. The inverse element

The condition for the existence of an inverse element reads: *To every element a there corresponds an element —a such that*

$$(-a) + a = a + (-a) = 0$$

is true.

Here again only the *existence* of the element $-a$ is asserted, and not its *uniqueness*. We establish this uniqueness in the following theorem.

Theorem.—If corresponding to a given element a there exists an element a', *which satisfies one of the conditions*

$$a + a' = 0 \quad \text{or} \quad a' + a = 0$$

then

$$a' = -a$$

Proof.—Let $a + a' = 0$. Then it follows that

$$(-a) + (a + a') = (-a) + 0 = -a$$

and therefore

$$[(-a) + a] + a' = -a$$

whence

$$0 + a' = -a$$

i.e.

$$a' = -a$$

In an analogous manner we can deduce $a' = -a$ from the assumption $a' + a = 0$.

Therefore corresponding to a given element a there exists exactly one element x which satisfies each of the equations $a + x = 0$, $x + a = 0$, namely the element $-a$.

Let us consider now the element $-a$. Since the element a satisfies the equation

$$-a + a = 0$$

it is just the element $x = -(-a)$ corresponding to $-a$ of which we were speaking. Therefore

$$-(-a) = a$$

4. Subtraction. The second rule for the removal of brackets

Suppose two elements a and b of the group G are given. Corresponding to the elements a and b there are inverse elements $-a$ and $-b$.

The sum of the elements b and $-a$ will be called the difference * *between the element b and the element a, and will be denoted by $b - a$, thus:*

$$b + (-a) = b - a \tag{1}$$

Hence this equation *defines the difference $b - a$*, i.e. defines subtraction as an operation by means of which the difference of the elements b and a is determined. On the basis of the associative law and the definition of the element $-a$ it follows that

$$(b - a) + a = b + (-a) + a = b + (-a + a) = b \tag{2}$$

The element b is therefore equal to the sum of the difference $b - a$ and the element a.†

In other words $b - a$ is a solution of the equation

$$x + a = b \tag{3}$$

It is also the only one; for if the element c is a solution of the equation (3), then $c + a = b$, which means that

$$c + a + (-a) = b + (-a)$$

* Sometimes " right difference ". See below.

† In non-commutative groups the sum $b + (a - b)$ is not in general equal to a. This is very important in group theory.

and therefore

$$c = b + (-a) = b - a$$

Similarly the equation

$$a + x = b \tag{4}$$

has the unique solution $-a + b$.

Remark.—Often we call the solution of the equation (3), viz. the element $b - a = b + (-a)$, the *right* difference, and the solution of the equation (4), viz. $-a + b$, the *left* difference of the elements b and a. For commutative groups evidently these two concepts of difference coincide.

Corollary.—If either $a + b = a + c$ *or* $b + a = c + a$ *then* $b = c$.

The main property of subtraction is expressed by the formula

$$-(a + b) = -b - a$$

[We remind the reader that in what follows $-b - a$ stands for $-b + (-a)$, i.e. for the sum of the two elements $-b$ and $-a$.]

Now the element $-(a + b)$ is the uniquely determined element x of the group, which satisfies the condition

$$a + b + x = 0 \tag{5}$$

But

$$a + b + [(-b) + (-a)] = a + [b + (-b)] + (-a)$$
$$= a + 0 + (-a) = a + (-a) = 0$$

Thus the element $x = -b - a$ satisfies the condition (5), and hence $-(a + b) = -b - a$.

By means of complete induction we obtain from this the general result

$$-(a_1 + \ldots + a_n) = -a_n - a_{n-1} - \ldots - a_1$$

where the element on the right-hand side stands for

$$(-a_n) + (-a_{n-1}) + \ldots + (-a_1)$$

From this, according to the definition of subtraction, it follows that

$$c - (a + b) = c - b - a$$

and in general

$$c - (a_1 + \ldots + a_n) = c - a_n - a_{n-1} - \ldots - a_1 \tag{6}$$

In commutative groups the order of the elements is irrelevant, and we can write

$$c - (a_1 + \ldots + a_n) = c - a_1 - \ldots - a_n \tag{6'}$$

Formula (1) in section 1 and formula (6′) express the familiar, rules of elementary algebra for the removal of brackets in addition and subtraction.

5. Remarks on the group axioms

We have not set ourselves the task of giving a *smallest* number of assumptions which are sufficient to define the concept of a group. We have postulated that the null element shall satisfy the conditions

$$a + 0 = 0 + a = a$$

and that to every element a shall correspond an inverse element $-a$ satisfying the conditions

$$a + (-a) = (-a) + a = 0$$

But it is in fact sufficient to assume only one of the conditions

$$a + 0 = a \quad \text{or} \quad 0 + a = a$$

and likewise only one of the conditions

$$a + (-a) = 0 \quad \text{or} \quad (-a) + a = 0$$

Finally we mention that in the definition of a group (§ 2) the axioms II and III, on the existence of a null element and of an element inverse to every given element, can be replaced by a single axiom, namely the following:

The existence of a difference of any two group elements: For any two elements a and b we can find x and y such that $a + x = b$ and $y + a = b$.

We leave it to the reader to justify these statements. (He may for example read up the proofs in *The Theory of Groups* by A. G. Kurosh.)

EXERCISES ON CHAPTER I

1. Show that in an addition table for a finite group every element can appear no more than once in each row or column, and hence that every element appears exactly once in each row or column. (For this proof you will need the inverse element.)

2. Make use of the property proved in Ex. 1 to show that there is only one possible addition table for a group of order 3.

3. Make use of this same property to write down the possible addition tables of groups of order 4. Deduce that every group of order 4 is Abelian.

4. If a is an element of a finite group of order r show that at least one of the elements

$$a,\ a + a,\ \ldots,\ \underbrace{a + \ldots + a}_{r \text{ times}}$$

is equal to the null element of the group.

5. Which of the following are groups?

(i) The positive real numbers (a) with respect to ordinary addition as the group operation, and (b) with respect to ordinary multiplication.*

(ii) The positive and negative whole numbers and 0 with respect to multiplication.

(iii) The even (odd) numbers with respect to addition (0 is even).

* See Chapter III for remarks on the additive and multiplicative terminology for a group.

Chapter II

GROUPS OF PERMUTATIONS

§ 1. Definition of a permutation group

If the three people David, John, and Peter are sitting, in this order from left to right, on a seat, then they can regroup themselves in six different ways, namely (always numbering from left to right):

(1) David, John, Peter
(2) David, Peter, John
(3) John, David, Peter
(4) John, Peter, David
(5) Peter, David, John
(6) Peter, John, David

The change from one seating-arrangement to another is called a *permutation*. We write permutations in the following way:

$$\begin{pmatrix} \text{David, John, Peter} \\ \text{John, Peter, David} \end{pmatrix}$$

and understand by this one that John has taken David's place, that Peter has taken John's, and that David has taken Peter's.

In a similar way we can speak of permutations of objects of any kind. Since the particular nature of the objects permuted is irrelevant here, we shall denote these objects by numbers, and speak of *permutations of numbers*. Thus with the three numbers 1, 2, 3 we can form the following permutations:

$$\begin{pmatrix} 1 & 2 & 3 \\ 1 & 2 & 3 \end{pmatrix} \quad \begin{pmatrix} 1 & 2 & 3 \\ 1 & 3 & 2 \end{pmatrix} \quad \begin{pmatrix} 1 & 2 & 3 \\ 2 & 1 & 3 \end{pmatrix} \quad \begin{pmatrix} 1 & 2 & 3 \\ 2 & 3 & 1 \end{pmatrix} \quad \begin{pmatrix} 1 & 2 & 3 \\ 3 & 1 & 2 \end{pmatrix} \quad \begin{pmatrix} 1 & 2 & 3 \\ 3 & 2 & 1 \end{pmatrix}$$

Each permutation signifies that the numbers standing in the top row are replaced respectively by those standing underneath them in the bottom row. We call the first permutation $\begin{pmatrix} 1 & 2 & 3 \\ 1 & 2 & 3 \end{pmatrix}$ the *identity*: here every number stays in its original place.

In the second permutation $\begin{pmatrix} 1 & 2 & 3 \\ 1 & 3 & 2 \end{pmatrix}$ the number 1 remains fixed, the number 3 takes the place of the number 2, and the number 2 takes the place of the number 3; similarly for the other permutations.

A permutation on n numbers 1, 2, . . . , n may be written in the general form

$$\begin{pmatrix} 1 & 2 & \dots & n \\ i_1 & i_2 & \dots & i_n \end{pmatrix}$$

Here $i_1, i_2, \dots, i_n$ are just the numbers $1, 2, \dots, n$ again, but written in a different order. By way of example we consider

$$\begin{pmatrix} 1 & 2 & 3 & 4 & 5 \\ 3 & 1 & 4 & 5 & 2 \end{pmatrix}$$

Evidently here $n = 5$, $i_1 = 3$, $i_2 = 1$, $i_3 = 4$, $i_4 = 5$, $i_5 = 2$.

It is well known that there are $n!$ possible permutations of n numbers.

We turn back to the consideration of permutations on three numbers. The *addition* of two permutations means their successive application, the first followed by the second. The result is again a permutation, and we call it the sum of the two given permutations.

By way of example we add the permutations

$$\begin{pmatrix} 1 & 2 & 3 \\ 2 & 1 & 3 \end{pmatrix} \quad \text{and} \quad \begin{pmatrix} 1 & 2 & 3 \\ 3 & 2 & 1 \end{pmatrix}$$

By the first permutation 1 is replaced by 2, by the second permutation 2 is unchanged, and therefore by applying the first permutation and then the second 1 is replaced by 2. In a similar way, by their successive application, 2 is replaced by 3, and 3 is replaced by 1. Therefore

$$\begin{pmatrix} 1 & 2 & 3 \\ 2 & 1 & 3 \end{pmatrix} + \begin{pmatrix} 1 & 2 & 3 \\ 3 & 2 & 1 \end{pmatrix} = \begin{pmatrix} 1 & 2 & 3 \\ 2 & 3 & 1 \end{pmatrix} \qquad (1)$$

In this way we can add together any two permutations. In order to be able to write down the result of all these additions in a convenient form, we introduce the following notation:

$$P_0 = \begin{pmatrix} 1 & 2 & 3 \\ 1 & 2 & 3 \end{pmatrix} \quad P_1 = \begin{pmatrix} 1 & 2 & 3 \\ 1 & 3 & 2 \end{pmatrix} \quad P_2 = \begin{pmatrix} 1 & 2 & 3 \\ 2 & 1 & 3 \end{pmatrix} \quad P_3 = \begin{pmatrix} 1 & 2 & 3 \\ 2 & 3 & 1 \end{pmatrix}$$

$$P_4 = \begin{pmatrix} 1 & 2 & 3 \\ 3 & 1 & 2 \end{pmatrix} \quad P_5 = \begin{pmatrix} 1 & 2 & 3 \\ 3 & 2 & 1 \end{pmatrix}$$

P_0 denotes the identical permutation.

We obtain the following *addition table*:

First Term	Second Term					
	P_0	P_1	P_2	P_3	P_4	P_5
P_0	P_0	P_1	P_2	P_3	P_4	P_5
P_1	P_1	P_0	P_3	P_2	P_5	P_4
P_2	P_2	P_4	P_0	P_5	P_1	P_3
P_3	P_3	P_5	P_1	P_4	P_0	P_2
P_4	P_4	P_2	P_5	P_0	P_3	P_1
P_5	P_5	P_3	P_4	P_1	P_2	P_0

In order to find the sum of two permutations, for example $P_2 + P_4$, we must take the row which is headed by the first permutation and the column which is headed by the second. The sum of the two permutations stands at the point of intersection of the selected row and column: $P_2 + P_4 = P_1$.

We carry out the details of the calculation:

$$P_2 = \begin{pmatrix} 1 & 2 & 3 \\ 2 & 1 & 3 \end{pmatrix} \qquad P_4 = \begin{pmatrix} 1 & 2 & 3 \\ 3 & 1 & 2 \end{pmatrix}$$

and, by considerations similar to those that led to equation (1), we have

$$\begin{pmatrix} 1 & 2 & 3 \\ 2 & 1 & 3 \end{pmatrix} + \begin{pmatrix} 1 & 2 & 3 \\ 3 & 1 & 2 \end{pmatrix} = \begin{pmatrix} 1 & 2 & 3 \\ 1 & 3 & 2 \end{pmatrix}$$

therefore

$$P_2 + P_4 = P_1$$

We leave it to the reader to verify the whole of the addition table in this way.

We convince ourselves at once that this addition satisfies the associative law.

The identical permutation $P_0 = \begin{pmatrix} 1 & 2 & 3 \\ 1 & 2 & 3 \end{pmatrix}$ clearly plays the role of a null element.

Finally every permutation has an inverse which when added to it results in the identical permutation. The permutation inverse to a given permutation brings back to their original places all the numbers altered by the given permutation. Thus for example

$$-\begin{pmatrix} 1 & 2 & 3 \\ 2 & 3 & 1 \end{pmatrix} = \begin{pmatrix} 1 & 2 & 3 \\ 3 & 1 & 2 \end{pmatrix}$$

In order to find, from the addition table, the inverse of a given permutation, we must look for the element P_0 in the row corresponding to the given permutation; the column in which P_0 lies corresponds to the required inverse. We easily verify that this gives:

$$\begin{array}{ll} -P_0 = P_0 & -P_3 = P_4 \\ -P_1 = P_1 & -P_4 = P_3 \\ -P_2 = P_2 & -P_5 = P_5 \end{array}$$

Therefore addition of permutations satisfies all the group axioms. The set of all permutations on three elements is therefore a group. We denote it by S_3. The group S_3 is finite, and of order 6. It is *not commutative.* In particular, for example:

$$\begin{array}{l} P_2 + P_3 = P_5 \\ P_3 + P_2 = P_1 \end{array}$$

§ 2. The concept of a subgroup

Examples from the theory of permutation groups

1. Examples and definition

It is natural to ask oneself the question: Is it possible to find a proper subset of the group of all permutations on three numbers, which is itself a group with respect to the same law of addition? We easily convince ourselves that it is possible.

Let us consider for example the two elements P_0 and P_1. We deduce from the addition table that

$$\begin{cases} P_0 + P_0 = P_0 \\ P_0 + P_1 = P_1 \\ P_1 + P_0 = P_1 \\ P_1 + P_1 = P_0 \end{cases}$$

We see that all the group axioms are satisfied, and in particular $-P_0 = P_0$ and $-P_1 = P_1$. This means that the two elements P_0 and P_1 form a group, which is a subset of the group of all permutations on three numbers.

In a similar way we convince ourselves that the two elements P_0 and P_2 form a group, as also do the elements P_0 and P_5.

The two elements P_0 and P_3 do not form a group, since

$$P_3 + P_3 = P_4$$

i.e. the sum of P_3 with itself is not one of the pair P_0, P_3; nor do the elements P_0 and P_4 form a group. These simple considerations justify the introduction of the following general definition:

If any group G is given, and if the set H, consisting of certain elements of G, is itself a group with respect to the same law of addition which holds in G, then H is called a subgroup of the group G. Thus each of the pairs (P_0, P_1), (P_0, P_2), (P_0, P_5) is a subgroup of order 2 of the group S_3. The group S_3 possesses no other subgroups of order 2. From the definition of a subgroup it follows that every subgroup H of a group G must contain the null element of the group G; therefore every subgroup of order 2 of the group S_3 has the form (P_0, P_i), where $i = 1, 2, 3, 4$ or 5. But we have seen that i cannot be equal to 3 or 4, and therefore there remain only the subgroups considered:

$$(P_0, P_1), \quad (P_0, P_2), \quad (P_0, P_5)$$

The group S_3 also possesses a subgroup consisting of three elements (a subgroup of order 3). This is the subgroup (P_0, P_3, P_4). The reader will be able to convince himself that this is the only subgroup of order 3 of S_3. There are no subgroups of orders 4 and 5 of the group S_3.*

Therefore the group S_3 has the following subgroups: three subgroups of order 2, namely (P_0, P_1), (P_0, P_2), (P_0, P_5); one subgroup of order 3, namely (P_0, P_3, P_4).

In the same way as we have investigated the group S_3, we can also investigate the group S_4, which consists of all permutations on four numbers.

The group S_4 is of order $1 . 2 . 3 . 4 = 24$.

In general for arbitrary n the permutations on n numbers form the group S_n of order $1 . 2 . 3 . \dots . n$.

The law of addition is the same in each of these groups: The addition of two permutations on n numbers means their successive application working from left to right.

* We can convince ourselves of this by investigating the ten subsets of the group S_3, which contain the element P_0 and consist of four elements, as well as the five subsets which contain P_0 and consist of five elements. But the non-existence of subgroups of S_3 of orders 4 and 5 follows at once from the following general theorem which will be proved later (Chapter VIII): *The order of every subgroup H of a finite group G is a divisor of the order of the group G.*

We remark finally that the group S_n of all permutations on n elements is often also called the *symmetric group* of permutations on n elements.

Every subgroup of the group S_n is called a *permutation group.*

2. A condition for a subset of a group to be a subgroup

In order to prove that a certain subset H of a group G is a subgroup, we make use of the following appropriate general theorem: *A subset H of a group G is a subgroup of G if and only if the following conditions are satisfied*:

1. *The sum of two elements a and b of H (in the sense in which addition is defined in G) is again an element of H.*
2. *The null element of the group G is an element of H.*
3. *The inverse of each element of H is again an element of H.*

In order to prove this it is sufficient to observe that our conditions simply require that the law of addition defined in G and limited to elements of H satisfies all the group axioms. We need not postulate the associative law. It is satisfied for the addition of arbitrary elements of the set G, and therefore in particular also when these elements belong to the set H.

§ 3.* Permutations considered as mappings of a finite set onto itself. Even and odd permutations

1. Permutations considered as mappings

We have investigated the concept of a permutation in an elementary and somewhat primitive way, as is usual. If we do not mind using general mathematical terminology, then we can define a permutation on n elements simply *as a one-to-one mapping f of the set of the given n elements onto itself.*

We assume that our elements are the numbers $1, 2, 3, \ldots, n$; then a permutation $\begin{pmatrix} 1 & 2 & 3 & \ldots n \\ a_1 & a_2 & a_3 & \ldots a_n \end{pmatrix}$ is specified by a function

$$a_k = f(k), \quad k = 1, 2, \ldots, n$$

whose argument and values are the numbers $1, 2, 3, \ldots, n$.

* The reader to whom this paragraph presents difficulties may omit it at a first reading, and need only come back to it just before Chapter VI. In any case, before reading this paragraph, the reader must be familiar with the whole of the appendix which is at the end of the book.

The values of the function which correspond to two different values of the argument are themselves always different.

In particular a permutation is completely determined when the value $f(k)$, i.e. a_k, is known for every k.

From this it follows that it is quite unimportant in what order we write down the numbers in the top row. What is important is that underneath the number k there stands the corresponding number a_k.

For example

$$\begin{pmatrix} 1 & 2 & 3 & 4 & 5 \\ 2 & 4 & 3 & 5 & 1 \end{pmatrix} \quad \text{and} \quad \begin{pmatrix} 3 & 4 & 5 & 2 & 1 \\ 3 & 5 & 1 & 4 & 2 \end{pmatrix}$$

represent two ways of writing one and the same permutation. This observation which is basically self-evident can also be formulated as follows: Suppose the permutation

$$A = \begin{pmatrix} 1 & 2 & 3 & \dots & n \\ a_1 & a_2 & a_3 & \dots & a_n \end{pmatrix} \tag{1}$$

is given.

If

$$P = \begin{pmatrix} 1 & 2 & 3 & \dots & n \\ p_1 & p_2 & p_3 & \dots & p_n \end{pmatrix} \tag{2}$$

is any permutation on the same numbers $1, 2, 3, \dots, n$, then the permutation (1) can also be written in the form

$$\begin{pmatrix} p_1 & p_2 & p_3 & \dots & p_n \\ a_{p_1} & a_{p_2} & a_{p_3} & \dots & a_{p_n} \end{pmatrix}$$

2. Even and odd permutations

Suppose the permutation

$$A = \begin{pmatrix} 1 & 2 & 3 & \dots & n \\ a_1 & a_2 & a_3 & \dots & a_n \end{pmatrix}$$

is given.

We consider an arbitrary set which consists of any two of the numbers $1, 2, 3, \dots, n$, and we denote these two numbers by i and k. This set is called a *number pair*. It is the pair consisting of the elements i and k and it is denoted by (i, k).* It is well known that the

* In this context the concept of a pair does not imply any condition on the order of the elements of the pair: (i, k) and (k, i) are two ways of writing one and the same pair. The pairs of elements which can be selected from n given elements are also called *combinations* of the second class of the n elements. The combinations of class p are just the subsets consisting of p elements.

number of pairs which can be chosen from n given elements is equal to*

$$\binom{n}{2} = \frac{n(n-1)}{1\,.\,2}$$

The pair consisting of the numbers i and k is called *regular* with respect to the permutation A if the differences $i - k$ and $a_i - a_k$ have the same sign. This means that if $i < k$ then we must have $a_i < a_k$, and if $i > k$ then $a_i > a_k$. Otherwise we say that the pair is *irregular* with respect to the permutation, or that it is an *inversion*. Thus if the pair (i, k) is an inversion then either $i < k$ and $a_i > a_k$ or $i > k$ and $a_i < a_k$.

By way of illustration we consider the permutations of the group S_3.

In the permutation $P_0 = \begin{pmatrix} 1 & 2 & 3 \\ 1 & 2 & 3 \end{pmatrix}$ there is no inversion; all pairs are regular.

In the permutation $P_1 = \begin{pmatrix} 1 & 2 & 3 \\ 1 & 3 & 2 \end{pmatrix}$ there is a single inversion (2, 3).

In the permutation $P_2 = \begin{pmatrix} 1 & 2 & 3 \\ 2 & 1 & 3 \end{pmatrix}$ there is a single inversion (1, 2).

In the permutation $P_3 = \begin{pmatrix} 1 & 2 & 3 \\ 2 & 3 & 1 \end{pmatrix}$ there are two inversions: (1, 3) and (2, 3).

In the permutation $P_4 = \begin{pmatrix} 1 & 2 & 3 \\ 3 & 1 & 2 \end{pmatrix}$ there are two inversions: (1, 2) and (1, 3).

In the permutation $P_5 = \begin{pmatrix} 1 & 2 & 3 \\ 3 & 2 & 1 \end{pmatrix}$ there are three inversions: (1, 2), (1, 3) and (2, 3).

Definition.—A permutation which contains an even number of inversions is called an *even* permutation; a permutation which contains an odd number of inversions is called an *odd* permutation.

We have seen that the even permutations P_0, P_3, and P_4 in the group S_3 form a subgroup. We now set ourselves the problem of showing that this is true for every group S_n.

The proof depends on a few preliminary considerations to which we now turn our attention.

By the sign of the permutation A we understand the number $+1$ if the permutation is even, and the number -1 if it is odd.

* We remark incidentally that the method given below allows us to avoid the logically unsound procedure which is often presented in school.

Departing from the usual terminology we shall understand here by the sign of the rational number r the number $+1$ when $r > 0$, the number -1 when $r < 0$, and the number 0 when $r = 0$.

We denote the sign of the number r by (sgn r),* and similarly the sign of the permutation A by (sgn A).

With this terminology it is clear that the sign of the permutation A is equal to the product of the signs of the $\frac{1}{2}n(n-1)$ numbers of the form $\dfrac{i-k}{a_i - a_k}$, where the fraction $\dfrac{i-k}{a_i - a_k} = \dfrac{k-i}{a_k - a_i}$, for each pair of numbers i, k taken from $1, 2, 3, \ldots, n$, is only formed once.

We use this observation in the proof of the following theorem:

The sign of the sum of two permutations is equal to the product of their signs.

Suppose we are given two permutations

$$A = \begin{pmatrix} 1 & 2 & 3 & \ldots & n \\ a_1 & a_2 & a_3 & \ldots & a_n \end{pmatrix} \qquad B = \begin{pmatrix} 1 & 2 & 3 & \ldots & n \\ b_1 & b_2 & b_3 & \ldots & b_n \end{pmatrix}$$

Their sum is evidently the permutation

$$A + B = \begin{pmatrix} 1 & 2 & 3 & \ldots & n \\ b_{a_1} & b_{a_2} & b_{a_3} & \ldots & b_{a_n} \end{pmatrix} \tag{1}$$

The sign of A is equal to the product of the signs of all the fractions

$$\frac{i-k}{a_i - a_k}$$

The sign of B is equal to the product of the signs of all the fractions

$$\frac{i-k}{b_i - b_k}$$

But since we may evidently also write

$$B = \begin{pmatrix} a_1 & a_2 & \ldots & a_n \\ b_{a_1} & b_{a_2} & \ldots & b_{a_n} \end{pmatrix}$$

it follows that:

The sign of B is equal to the product of the signs of all the fractions $\dfrac{a_i - a_k}{b_{a_i} - b_{a_k}}$. From this it follows immediately that

* Evidently for any two rational numbers r, s we have (sgn r)(sgn s) = (sgn rs). We make use of this result below.

$$(\operatorname{sgn} A)\,.\,(\operatorname{sgn} B) = \text{the product of all}\left(\operatorname{sgn}\frac{i-k}{a_i-a_k}\right)\cdot\left(\operatorname{sgn}\frac{a_i-a_k}{b_{a_i}-b_{a_k}}\right)$$

$$= \text{the product of all}\left(\operatorname{sgn}\frac{i-k}{a_i-a_k}\cdot\frac{a_i-a_k}{b_{a_i}-b_{a_k}}\right)$$

$$= \text{the product of all}\left(\operatorname{sgn}\frac{i-k}{b_a-b_{a_k}}\right)$$

But this last product is the sign of the permutation

$$\begin{pmatrix}1 & 2 & 3 & \dots n \\ b_{a_1} & b_{a_2} & b_{a_3} & \dots b_{a_n}\end{pmatrix}$$

and therefore of the permutation $A + B$; and this is what we set out to prove.

From the theorem just proved it follows that: *The sum of two like* permutations is even and the sum of two unlike† permutations is odd.* The identical permutation contains no inversion and is therefore an even permutation. Further

$$A + (-A) = 0$$

and therefore the sum of a given permutation and its inverse is even. From this it follows, by what we have just proved, that a permutation and its inverse are like (i.e. belong to the same class).

Therefore: *The sum of two even permutations is even, the identical permutation is even, and the inverse of an even permutation is even.*

It follows from this that the set of all even permutations on n elements is a subgroup of the group S_n of all permutations on n elements. The group of all even permutations on n elements is called *the alternating group on n elements, and is denoted by A_n.*

Theorem.—The order of the group A_n is equal to $\frac{1}{2}n!$. In other words, just half of the permutations on n elements belong to the group A_n. In order to convince ourselves of this it is sufficient to establish a one-to-one correspondence between the set of all even permutations and the set of all odd permutations on n elements. We establish this correspondence if we choose any odd permutation P, and then

* That is to say, the sum of two even or of two odd permutations.

† That is, the sum of an even and an odd permutation, or the sum of an odd and an even permutation.

associate with every even permutation A the permutation $P + A$. In this way it results that:

1. To every even permutation there corresponds an odd permutation.
2. To two different even permutations there correspond two different odd permutations.
3. Every odd permutation B is associated with one (and only one) even permutation, namely the permutation $-P + B$.

Therefore there is a one-to-one correspondence between the set of all even permutations and the set of all odd permutations.

EXERCISES ON CHAPTER II

1. (*a*) Find the sum $\begin{pmatrix} 1 & 2 & 3 & 4 \\ 1 & 3 & 4 & 2 \end{pmatrix} + \begin{pmatrix} 1 & 2 & 3 & 4 \\ 2 & 3 & 1 & 4 \end{pmatrix}$.

(*b*) Calculate the inverse of each of the permutations in (*a*).

(*c*) Find the number of inversions in each of the permutations in (*a*).

2. How many times must $\begin{pmatrix} 1 & 2 & 3 & 4 \\ 1 & 3 & 4 & 2 \end{pmatrix}$ be added to itself to produce $\begin{pmatrix} 1 & 2 & 3 & 4 \\ 1 & 2 & 3 & 4 \end{pmatrix}$?

3. (*a*) Find some subgroups of the group of all real numbers (where the group operation is ordinary addition).

(*b*) List the subgroups of Klein's four-group.

(*c*) List the subgroups of the group of rotations of a square.

4. Prove that the permutations $\begin{pmatrix} 1 & 2 & 3 & 4 \\ a_1 & a_2 & a_3 & a_4 \end{pmatrix}$ leaving invariant the polynomial $x_1x_2 + x_3 + x_4$, i.e. for which $x_{a_1}x_{a_2} + x_{a_3} + x_{a_4}$ is identical with $x_1x_2 + x_3 + x_4$, form a subgroup H of order 4 of the symmetric group S_4, and write down its addition table.

(H is called the group of the polynomial $x_1x_2 + x_3 + x_4$. A polynomial in x_1, x_2, x_3, x_4 whose group is S_4 itself is called symmetrical.)

5. Find the group of the polynomial $x_1x_2 + x_3x_4$, and verify that it contains as a subgroup the group H of Ex. 4.

Chapter III

SOME GENERAL REMARKS ABOUT GROUPS
THE CONCEPT OF ISOMORPHISM

§ 1. The " additive " and the " multiplicative " terminology in group theory

The principal ingredients in the group concept are:

(*a*) The set of objects (numbers, permutations, rotations, etc.) which form the elements of the group.

(*b*) A certain *operation* or *law of combination*, which we have called addition and which, given any two elements a and b of our group, allows us to find a third element $a + b$ of this group.

We have chosen the word addition to describe the law of combination in our group. Obviously the choice of this or of any other word is basically of no significance. We could just as well speak of the *multiplication* of the elements of any group instead of their addition, provided we use not the *additive* but the *multiplicative* terminology. We are already familiar with the additive terminology, or additive notation, for a group. Now let us consider how the group axioms can be expressed in multiplicative terminology.

First of all we postulate that, corresponding to any two elements a and b of our set G (see Chapter I, § 2), there is a uniquely defined element $a \,.\, b$, the product of the two elements a and b.

The group axioms themselves then take the following form.

I. *The Associative Law*:

$$(ab)c = a(bc)$$

II. *The condition for the existence of a unit element. Among the elements of G there is a uniquely determined element which we call the unit element and denote by e, and which is such that*

$$ae = a = ea$$

for an arbitrary choice of the element a.

III. *The condition for the existence of an inverse of each given element. Corresponding to any given element a of the set G we can find an element a^{-1} of G such that*

$$aa^{-1} = e = a^{-1}a$$

We see that if the law of combination defined in a given group and originally called addition is now conceived of as multiplication, then it is sensible to call the null element a unit, and instead of denoting the inverse of a by $-a$ to denote it instead by a^{-1}.

Historically the "multiplicative" terminology came first; nowadays it is used by nearly all authors. In some cases the additive terminology is preferable, in other cases the multiplicative terminology. Finally there are cases when both are equally suitable.

An example in which the additive terminology is naturally the more suitable is provided by the group of whole numbers: The group operation here is just ordinary arithmetical addition, the null element 0 is the ordinary zero; the element $-a$ has its ordinary arithmetical meaning of "minus a".

We can object that it is unnatural and inconvenient to call ordinary arithmetical addition by the name of multiplication, and to speak of the zero as the unit, and so on. But the reader must be quite clear that this terminology, in spite of all its inconveniences, is perfectly possible, and at any rate leads to no kind of inconsistency so long as we restrict ourselves to the study only of the *group* of whole numbers, that is to say that we consider only a *single operation* on whole numbers, namely arithmetical addition. If we should wish to consider, as well as arithmetical addition, multiplication also, using the word in the ordinary arithmetical sense, then calling addition multiplication, in the way that we have been describing, would naturally be utterly confusing.

As an example of a group for which on the other hand the multiplicative terminology is more suitable, we consider the group R consisting of all positive and negative rational numbers,* that is, of all rational numbers *different from zero*. We consider ordinary arithmetical multiplication as the group operation in R. It is known to be associative. The number 1 satisfies axiom II with respect to this operation:

$$a \,.\, 1 = a \text{ for any } a$$

Finally, corresponding to any element of the set R (therefore to any rational number $a \neq 0$) there is a rational number $a^{-1} = 1/a \neq 0$,

* By rational numbers we understand all whole numbers as well as all fractions p/q (p, q integral, $q \neq 0$).

which satisfies the condition $a \cdot a^{-1} = 1$. Therefore all the group axioms are satisfied, i.e. the rational numbers different from zero form a group with respect to arithmetical multiplication. Since $ab = ba$ this group is *commutative*. It contains as a subgroup the group of all positive rational numbers ($a > 0$). In these groups we naturally use the multiplicative terminology.

The reader may convince himself that the negative rational numbers do not form a group with respect to ordinary arithmetical multiplication.

Also the set of all rational numbers (including the zero) does not form a group with respect to arithmetical multiplication, since the zero possesses no inverse number. On the other hand we see at once that the set of all rational numbers forms a group with respect to arithmetical addition. The group of whole numbers is contained in this group as a subgroup.

Finally, on this question of terminology, we remark that for the permutation groups there is no serious ground for preferring the additive to the multiplicative terminology or the other way round. In the multiplicative terminology, however, one of the theorems of the previous chapter may be stated in a symmetrical form, namely: The sign of the product of two permutations is equal to the product of their signs.

Nowadays it is becoming more and more usual to change to the additive terminology when dealing with commutative groups; but we became acquainted with an exception to this rule when we spoke of the group of rational numbers different from zero. In this book we shall retain the additive terminology also for non-commutative groups.

§ 2. Isomorphic groups

We consider on the one hand the rotation group R_3 of an equilateral triangle (Chapter I, § 1) and on the other hand the subgroup A_3, consisting of the three elements P_0, P_3, P_4 (Chapter II, § 2), of the group of all permutations on three numbers. We denote by a_0, a_1, a_2 the elements of the group R_3. We now set up between the elements of the group R_3 and the elements of the group A_3 the following one-to-one correspondence:

$$a_0 \longleftrightarrow P_0$$

$$a_1 \longleftrightarrow P_3$$

$$a_2 \longleftrightarrow P_4$$

This correspondence *preserves addition* in the following sense: If an element in R_3 is written as the sum of two elements of R_3, thus for example $a_0 + a_1 = a_1$, $a_1 + a_1 = a_2$, $a_1 + a_2 = a_0$, and if we replace every element in the equation considered by the corresponding element from A_3 then the equation remains true.

We recognize that the groups R_3 and A_3, although they consist of elements of different natures (the one group consists of rotations of a triangle and the other of permutations of numbers), *have the same structure*: the addition tables of these groups differ from one another only in notation. If we change the notation we can therefore so name the elements that we obtain identical group tables.

Groups, whose addition tables become identical when the elements are suitably named, are called isomorphic groups.

The definition of isomorphism is usually worded somewhat differently. The process of suitably " naming " the elements in the addition table, which we speak of in this definition, means essentially that there is set up a one-to-one correspondence between the elements of the two groups. We give a definition of isomorphism on these lines, in terms of the idea of a one-to-one mapping.

Definition I.—Suppose that there is given a one-to-one correspondence

$$g \longleftrightarrow g'$$

between the set of all elements of the group G and the set of all elements of the group G'. We shall say that this correspondence is an *isomorphic correspondence* (or an *isomorphism*) between the two groups if it *preserves addition*. This means that:

If any relation of the form

$$g_1 + g_2 = g_3$$

holds between the elements of a group G, then the relation which is obtained by replacing the elements g_1, g_2, g_3 of the group G by the elements g_1', g_2', g_3' *which correspond to them* in the group G', namely

$$g_1' + g_2' = g_3'$$

is also valid.

Definition II.—Two groups are called isomorphic if it is possible to set up an isomorphic correspondence between them.

Remark. If we postulate that a relation

$$g_1 + g_2 = g_3$$

in the group G always implies a relation

$$g_1' + g_2' = g_3'$$

between the elements of the group G' which correspond to the elements g_1, g_2, g_3, then the converse is also true, namely: If for any three elements g_1', g_2', g_3' of the group G' we have the relation

$$g_1' + g_2' = g_3'$$

then the relation

$$g_1 + g_2 = g_3 \tag{1}$$

between the elements g_1, g_2, g_3 of the group G which correspond to the elements g_1', g_2', g_3' is also true. For if the relation (1) were not true then we should have

$$g_1 + g_2 = g_4 \neq g_3$$

On account of the one-to-one correspondence between G and G' there would correspond to the element g_4 of the group G an element g_4' in the group G', different from g_3'; but by hypothesis we deduce from

$$g_1 + g_2 = g_4$$

the equation $$g_1' + g_2' = g_4'$$

which contradicts $$g_1' + g_2' = g_3'$$

Theorem.—In the isomorphic mapping

$$g \longleftrightarrow g'$$

of the group G onto the group G' the null element of the one group corresponds to the null element of the other group. Every pair of inverse elements of the one group corresponds to a pair of inverse elements of the other group.

Suppose that g_0 is the null element of the group G, and that the element g_0' of the group G' corresponds to it in the given isomorphic correspondence between the groups G and G'. We prove that g_0' is the null element of the group G'. Since g_0 is the null element of the group G it follows that, for every element g of this group,

$$g + g_0 = g$$

On account of the isomorphic mapping $g \longleftrightarrow g'$ it is true that

$$g' + g_0' = g'$$

whence g_0' is the null element of the group G'.

Let g_1 and g_2 be a pair of inverse elements of the group G, so that

$$g_1 + g_2 = g_0$$

(where as above g_0 is the null element of the group G).
From this it follows that

$$g_1' + g_2' = g_0'$$

Since g_0' is the null element of the group G' then g_1' and g_2' are inverse elements of G'.

§ 3. Cayley's theorem *

We conclude this chapter by proving the following theorem, which was discovered by Cayley.†

Theorem.—Every finite group is isomorphic to a certain group of permutations.

Proof.—Let G be a finite group, n its order,

$$a_1, a_2, \ldots, a_n$$

its elements, and among these let a_1 be the null element.

We write out the elements

$$a_1 + a_i,\ a_2 + a_i,\ \ldots,\ a_n + a_i$$

for every $i = 1, 2, 3, \ldots, n$. For fixed i these elements are always distinct, and there are n of them; therefore they are always just the same elements $a_1, a_2, \ldots, a_n$, but taken in a different order. Write

$$a_1 + a_i = a_{i_1},\quad a_2 + a_i = a_{i_2},\ \ldots,\ a_n + a_i = a_{i_n}$$

Therefore there corresponds to the element a_i the permutation

$$P_i = \begin{pmatrix} a_1 & a_2 & \ldots & a_n \\ a_1 + a_i & a_2 + a_i & \ldots & a_n + a_i \end{pmatrix} = \begin{pmatrix} a_1 & a_2 & \ldots & a_n \\ a_{i_1} & a_{i_2} & \ldots & a_{i_n} \end{pmatrix}$$

or also the permutation

$$P_i' = \begin{pmatrix} 1 & 2 & \ldots & n \\ i_1 & i_2 & \ldots & i_n \end{pmatrix}$$

which only differs from the permutation P_i in that in P_i it is the

* The reader who has passed over § 3 of the previous chapter must also pass over this paragraph.

† The English mathematician Cayley (1821–1895) was one of the originators of the theory of groups.

elements of the group G itself which are permuted while in P_i' it is the uniquely determined indices of these elements.

For $i \neq k$, i.e. $a_i \neq a_k$, we also have $P_i \neq P_k$; for underneath the element a_1 in the permutation P_i there stands the element $a_1 + a_i = a_i$ while in the permutation P_k there stands $a_1 + a_k = a_k$.

We have therefore set up a one-to-one correspondence between the elements $a_1, a_2, \ldots, a_n$ of the group G and the permutations $P_1, P_2, \ldots, P_n$.

We must now prove, firstly that the permutations $P_1, P_2, \ldots, P_n$, with respect to the operation of addition of permutations, form a group, and secondly that this group is isomorphic to the group G.

First of all we observe that:

I. *Among the permutations $P_1, P_2, \ldots, P_n$ there is included the identity.*

Indeed since by hypothesis a_1 is the null element of the group G it follows that the permutation

$$P_1 = \begin{pmatrix} a_1 & a_2 & \ldots & a_n \\ a_1 + a_1 & a_2 + a_1 & \ldots & a_n + a_1 \end{pmatrix}$$

is the identical permutation.

Further we prove that: If $a_h = a_i + a_k$, then also $P_h = P_i + P_k$. Firstly we remark that

$$\begin{pmatrix} a_1 & a_2 & \ldots & a_n \\ a_1 + a_k & a_2 + a_k & \ldots & a_n + a_k \end{pmatrix}$$

and
$$\begin{pmatrix} a_1 + a_i & a_2 + a_i & \ldots & a_n + a_i \\ a_1 + a_i + a_k & a_2 + a_i + a_k & \ldots & a_n + a_i + a_k \end{pmatrix}$$

are two ways of writing one and the same permutation P_k. Both ways show that to each element a of the group G corresponds the element $a + a_k$ of the same group.

Thus we may write

$$P_k = \begin{pmatrix} a_1 + a_i & a_2 + a_i & \ldots & a_n + a_i \\ a_1 + a_i + a_k & a_2 + a_i + a_k & \ldots & a_n + a_i + a_k \end{pmatrix}$$

From this we see that the permutation

$$P_i + P_k = \begin{pmatrix} a_1 & a_2 & \ldots & a_n \\ a_1 + a_i & a_2 + a_i & \ldots & a_n + a_i \end{pmatrix}$$
$$+ \begin{pmatrix} a_1 + a_i & a_2 + a_i & \ldots & a_n + a_i \\ a_1 + a_i + a_k & a_2 + a_i + a_k & \ldots & a_n + a_i + a_k \end{pmatrix}$$

according to the general definition of addition of permutations, is identical with the permutation

$$\begin{pmatrix} a_1 & a_2 & \dots & a_n \\ a_1 + a_i + a_k & a_2 + a_i + a_k & \dots & a_n + a_i + a_k \end{pmatrix}$$

Since $a_i + a_k = a_h$, we have

$$\begin{pmatrix} a_1 & a_2 & \dots & a_n \\ a_1 + a_i + a_k & a_2 + a_i + a_k & \dots & a_n + a_i + a_k \end{pmatrix} = P_h$$

i.e.
$$P_i + P_k = P_h$$

This result may be stated in the following terms:

IIa. *To the sum of two elements of the group G corresponds the sum of the permutations corresponding to these elements.*

From this follows:

IIb. *The sum of any two permutations from the set* $P_1, P_2, \dots, P_n$ *is again one of the permutations* $P_1, P_2, \dots, P_n$.

We consider the permutation P_i, the element a_i and the element $-a_i = a_k$. Since $a_i + a_k = a_1$ then, by what we have just proved, $P_i + P_k = P_1$; but as we have seen P_1 is the identical permutation, and therefore $P_k = -P_i$.

Therefore:

III. *For arbitrary* $i = 1, 2, \dots, n$, $-P_i$ *is one of the permutations* $P_1, P_2, \dots, P_n$.

From IIb, I, and III it follows that the set of permutations $P_1, P_2, \dots, P_n$ is a group with respect to the usual definition of addition of permutations.

From IIa it follows that this group is isomorphic to the group G. Therefore Cayley's theorem is proved.

EXERCISES ON CHAPTER III

1. Prove that the group, which consists of the two elements a_0 and a_1 with the addition table

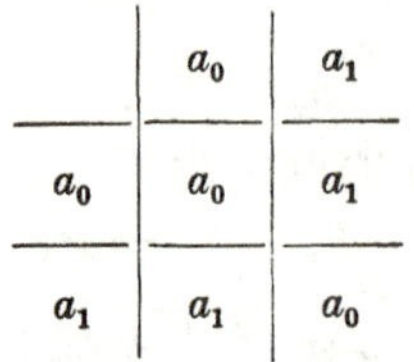

	a_0	a_1
a_0	a_0	a_1
a_1	a_1	a_0

is isomorphic to the group of rotations of an interval of a straight line (about its mid-point).

2. Prove that all groups of order 2 are isomorphic to one another.

3. Prove that all groups of order 3 are isomorphic to one another (see Chapter 1, Ex. 2).

4. Prove that every group of order 4 is isomorphic either to Klein's four-group or to the rotation group of a square. (These two groups are not isomorphic to one another.)

5. Prove that the group of all positive numbers (with arithmetical multiplication as the group operation) is isomorphic to the group of all real numbers (with arithmetical addition as the group operation). (The isomorphic mapping is set up by taking logarithms.)

6. If a group is isomorphic to one of its proper subgroups, what is its order?

7. Find a group of permutations on four numbers which is isomorphic (*a*) to the group of rotations of a square, and (*b*) to Klein's four-group.

8. Find a group of permutations on six numbers which is isomorphic to the symmetric group S_3.

Chapter IV

CYCLIC SUBGROUPS OF A GIVEN GROUP

§ 1. The subgroup generated by a given element of a given group

Let a be an arbitrary element of a group G. We add it to itself, thus forming the element $a + a$. This element we denote by $2a$. It must be stressed that $2a$ is only a *way of writing* the element $a + a$; on no account are we speaking here of the multiplication by 2 of the element a. Similarly we denote $a + a + a$ by $3a$, and in general we put

$$\underbrace{a + a + \ldots + a}_{n \text{ times}} = na$$

Further we consider the element $-a$ and denote in turn

$$\begin{array}{rcl} (-a) + (-a) & \text{by} & -2a \\ (-a) + (-a) + (-a) & \text{by} & -3a \\ \ldots\ldots & & \ldots \\ \underbrace{(-a) + (-a) + \ldots + (-a)}_{n \text{ times}} & \text{by} & -na \end{array}$$

This notation is justified by the fact that

$$na + (-na) = 0$$

In order to prove this assertion we first of all remark that it is clearly true in the case $n = 1$ (this follows from the definition of $-a$).* We assume that it is true for $n - 1$, and prove under this hypothesis that it is true for n. We have

$$\begin{aligned} na + (-na) &= [a + (n-1)a] + [-(n-1)a + (-a)] \\ &= a + \{(n-1)a + [-(n-1)a]\} + (-a) \end{aligned}$$

* On the understanding that $1a = a$ and $-1a = -a$.

But by hypothesis the expression in curly brackets is equal to zero, therefore

$$na + (-na) = a + 0 + (-a) = a + (-a) = 0$$

which is what we set out to prove.

We have defined the expression na for arbitrary positive and for arbitrary negative integral n. Finally we agree to write $0a = 0$ (where 0 on the left-hand side denotes the number zero and 0 on the right-hand side denotes the null element of the group).

Now let p and q be any two whole numbers. From our definition it follows that

$$pa + qa = (p + q)a$$

We obtain the following result:

The set $H(a)$ of the elements of a group G, which can be represented in the form na for integral n, form a group with respect to the law of addition defined in the group G.

Indeed we have:

1. The sum of two elements belonging to $H(a)$ is again an element of $H(a)$.

2. The null element belongs to $H(a)$.

3. To every element ma of $H(a)$ corresponds an element $-ma$ which likewise belongs to $H(a)$.

Therefore $H(a)$ is a subgroup of G. We call this subgroup *the subgroup of the group G generated by the element a.*

§ 2. Finite and infinite cyclic groups

We have defined the group $H(a)$ as the group consisting of all those elements of G which are representable in the form ma. But we have not yet considered the following question: Do two expressions m_1a and m_2a involving different integers m_1 and m_2 always give rise to two different elements of the group G, or can it happen that $m_1a = m_2a$ with m_1 and m_2 distinct?

We will concern ourselves with this problem now. Suppose there exist two different whole numbers m_1 and m_2 for which $m_1a = m_2a$. If we add the element $-m_1a$ to both sides of this last equation then we obtain

$$0 = (m_2 - m_1)a$$

Hence there exists a whole number m such that

$$ma = 0$$

Since from $ma = 0$ it follows that $-ma = 0$, we may always assume that the number m in the equation is positive.

We now select, from among all the natural numbers satisfying the condition $ma = 0$, the smallest one, and denote it by α. We have

$$a \neq 0, \ 2a \neq 0, \ \ldots, \ (\alpha - 1)a \neq 0; \ \alpha a = 0$$

We prove that all the elements

$$0 = 0a, \ a, \ 2a, \ \ldots, \ (\alpha - 1)a \tag{1}$$

are different from one another. Indeed if it were true that

$$pa = qa, \quad \text{with} \quad 0 \leqslant p < q \leqslant \alpha - 1$$

then, if we added $-pa$ to both sides of the last equation, we should obtain the result

$$(q - p)a = 0$$

But this would contradict the definition of the number α, since under our conditions we have

$$0 < q - p \leqslant \alpha - 1$$

Therefore all the elements (1) are different from one another. We prove that the whole group $H(a)$ is exhausted by the elements (1), so that therefore for arbitrary integral m we have

$$ma = ra, \quad \text{with} \quad 0 \leqslant r \leqslant \alpha - 1$$

To this end we divide m by α and represent m in the form

$$m = q\alpha + r \tag{2}$$

where q is the quotient and r is the remainder, satisfying the condition *

$$0 \leqslant r < \alpha$$

Then we have $$ma = (q\alpha + r)a = q\alpha \,.\, a + ra$$

and since $$q\alpha \,.\, a = q(\alpha a) = q \,.\, 0 = 0$$

also $$ma = ra$$

Therefore if there exist two numbers m_1 and m_2 such that $m_1 a = m_2 a$, then there exists a natural number α such that the group $H(a)$ is exhausted by the α mutually distinct elements

$$0,\ a,\ 2a,\ \ldots,\ (\alpha - 1)a \tag{1}$$

also $\alpha a = 0$, and more generally: The whole series

$$\ldots, -ma, \ldots, -a, 0, a, \ldots, ma, \ldots$$

simply consists of infinitely many repetitions, to the left and to the right, of the series (1). Indeed we have:

$$(\alpha + 1)a = \alpha a + a = a$$
$$(\alpha + 2)a = \alpha a + 2a = 2a$$
$$\cdots\cdots\cdots\cdots$$
$$(2\alpha - 1)a = a + (\alpha - 1)a = (\alpha - 1)a$$
$$2\alpha a = 0$$
$$(2\alpha + 1)a = a, \quad \text{and so on;}$$

and similarly in the left half:

$$-a = \alpha a - a = (\alpha - 1)a$$
$$-2a = \alpha a - 2a = (\alpha - 2)a$$
$$\cdots\cdots\cdots\cdots$$
$$-(\alpha - 1)a = \alpha a - (\alpha - 1)a = a$$
$$-\alpha a = 0, \quad \text{and so on.}$$

* Even for negative m the remainder r on division by $\alpha > 0$ is to be taken to be non-negative. Indeed if m is negative then $-m$ is positive and can be written in the form

$$-m = q'\alpha + r'$$

where q' and r' are non-negative. For $r' > 0$ we have

$$m = -q'\alpha - r' = -(q' + 1)\alpha + (\alpha - r')$$

We agree to call the number $-(q' + 1)$ the quotient and the *positive* number $\alpha - r' < \alpha$ the remainder when the negative number m is divided by the positive number α. For further details see Chapter VII, § 2.

In order to find the element of the group $H(a)$ which is formed by the sum

$$\underbrace{a + a + \ldots + a}_{m \text{ times}} = ma$$

or

$$\underbrace{(-a) + (-a) + \ldots + (-a)}_{m \text{ times}} = -ma$$

we must divide m or $-m$ by α. The non-negative remainder r which we obtain after this division satisfies the condition $0 \leqslant r \leqslant \alpha - 1$ and gives

$$ma = ra$$

From this it is also clear how the elements of the group $H(a)$ are to be added together; indeed

$$pa + qa = (p + q)a = ra$$

where r is the remainder when $p + q$ is divided by α.

We consider now a regular polygon of α sides. The angle at the centre subtended by a side of the polygon is equal to

$$\varphi = 2\pi/\alpha$$

The polygon is brought into coincidence with itself by rotations through angles 0 (the identical or zero rotation), φ, 2φ, . . . , $(\alpha - 1)\varphi$. If we identify rotations which only differ from one another by a whole number of complete revolutions, then these multiples of φ represent the only rotations bringing the polygon into coincidence with itself. The sum of the rotations through the angles $p\varphi$ and $q\varphi$ is evidently equal to the rotation through the angle $r\varphi$, where r is the remainder when $p + q$ is divided by α.

We see that: If we associate with the element ma of the group $H(a)$ the rotation of the polygon through the angle $m\varphi$, then we obtain an isomorphic mapping of the group $H(a)$ onto the group of rotations of the regular α-sided polygon.

*Groups which are isomorphic to groups of rotations of regular polygons are called finite cyclic groups.**

If therefore $m_1a = m_2a$ for certain m_1 and m_2 then the group $H(a)$ is a finite cyclic group.

The addition tables of cyclic groups of orders 3 and 4 were written

* We regard an interval of a straight line as a polygon with two sides and two vertices (see also Chapter V, § 3, section 3). The group consisting of the null element 0 alone is included as a trivial cyclic group of order 1 generated by 0.

down in Chapter I, § 1 (first and third examples). The addition table of a cyclic group of order α has the form:

	a_0	a_1	a_2	a_3	...	$a_{\alpha-1}$
a_0	a_0	a_1	a_2	a_3	...	$a_{\alpha-1}$
a_1	a_1	a_2	a_3	a_4	...	a_0
a_2	a_2	a_3	a_4	a_5	...	a_1
a_3	a_3	a_4	a_5	a_6	...	a_2
⋮	⋮	⋮	⋮	⋮	⋮	⋮
$a_{\alpha-3}$	$a_{\alpha-3}$	$a_{\alpha-2}$	$a_{\alpha-1}$	a_0	...	$a_{\alpha-4}$
$a_{\alpha-2}$	$a_{\alpha-2}$	$a_{\alpha-1}$	a_0	a_1	...	$a_{\alpha-3}$
$a_{\alpha-1}$	$a_{\alpha-1}$	a_0	a_1	a_2	...	$a_{\alpha-2}$

We can interpret this addition table as a second definition of a cyclic group of order α.

We have investigated the case that for a given element a of the group G there exist two different whole numbers m_1 and m_2 with the property that $m_1a = m_2a$.

We consider now the case that no two such numbers exist, so that therefore all the elements

$$\begin{aligned} \ldots, -ma, -(m-1)a, \ldots, -3a, -2a, \\ -a, 0, a, 2a, 3a, \ldots, ma, \ldots \qquad (3) \end{aligned}$$

are distinct. In this case there is a one-to-one correspondence between the elements (3) and the whole numbers: To the element ma corresponds the whole number m, and conversely. And with

$$m_1a + m_2a = m_3a$$

we also have

$$m_1 + m_2 = m_3$$

This one-to-one correspondence is therefore an isomorphism between the subgroup $H(a)$ and the group of all whole numbers.

Groups which are isomorphic to the group of all whole numbers are called infinite cyclic groups.

Further, since two groups A and B which are isomorphic to one and the same group C are evidently isomorphic to each other, it follows that all infinite cyclic groups are isomorphic to one another. Likewise all finite cyclic groups of the same order are isomorphic to one another.

We summarize the results of this paragraph.

Theorem.—Every element a of a group G generates a finite or infinite cyclic group $H(a)$. The order of the group $H(a)$ is called *the order of the element a.*

Finally we may also define finite or infinite cyclic groups as follows: *A group is called cyclic if it is generated by one of its elements.*

§ 3. Systems of generators

We now turn back to the cyclic group $H(a)$ which is generated by the element a of the group G. The element a generates the group $H(a)$ in the sense that every one of the elements of $H(a)$ is a sum of terms each of which is either equal to a or to $-a$.

The statement that " the element a generates the group $H(a)$ " is equivalent to the statement that " the element a is a *generating element* of the group $H(a)$ ".

However not every group is cyclic, i.e. not every group is generated by a single element. Non-cyclic groups are generated not by one but by many, sometimes by infinitely many, elements. The concept of a generating element leads to the concept of *a system of generators.*

*Definition.—A set E of elements of a group G is called a system of generators of this group if every element of the group is the sum of a finite number of terms, each of which is either an element of E or is the inverse of an element of E.**

Example.—We consider the plane with a Cartesian system chosen in it, and denote by G the set of points $P = (x, y)$ whose two coordinates x and y are whole numbers. We lay down the following rule of addition for points: The sum of the two points $P_1 = (x_1, y_1)$ and $P_2 = (x_2, y_2)$ is the point $P_3 = (x_3, y_3)$ with the coordinates $x_3 = x_1 + x_2$ and $y_3 = y_1 + y_2$. We see at once that with addition so

* Evidently the set of all the elements of any group is a (trivial) system of generators of this group. Therefore *every group possesses a system of generators.*

defined the set G is an Abelian group (see Chapter I, § 2, IV), and that the points (0, 1) and (1, 0) form a system of generators of this group.

Remark.—If the reader is familiar with the concept of a complex number, then he can easily show that the group just constructed is isomorphic to the group of complex numbers $x + iy$ with integral real and imaginary parts x and y (with addition as the group operation).

EXERCISES ON CHAPTER IV

1. Find the cyclic subgroups of the symmetric group S_3.

2. Show that an infinite cyclic group has an unlimited number of infinite cyclic subgroups, and that each of these subgroups is isomorphic to the original group.

3. Can an infinite cyclic group have a finite subgroup? (See Ex. 4 of Chapter I.)

4. Show that all subgroups of a cyclic group are themselves cyclic, and hence in particular that an infinite cyclic group is isomorphic to each of its subgroups.

5. Prove that a cyclic group of order m with the elements $0, a, 2a, \ldots, (m - 1)a$ is generated by the element ra provided that the greatest common divisor (g. c. d.) of m and r is equal to 1. (In particular a cyclic group of prime order is generated by any one of its elements.)

6. Prove that S_3 is generated by the two permutations $\begin{pmatrix} 1 & 2 & 3 \\ 2 & 1 & 3 \end{pmatrix}$, $\begin{pmatrix} 1 & 2 & 3 \\ 3 & 2 & 1 \end{pmatrix}$, but not by $\begin{pmatrix} 1 & 2 & 3 \\ 2 & 3 & 1 \end{pmatrix}$, $\begin{pmatrix} 1 & 2 & 3 \\ 3 & 1 & 2 \end{pmatrix}$.

7. Prove that every set of natural numbers whose g. c. d. is equal to 1 is a system of generators of the group of all whole numbers.

Chapter V

SIMPLE GROUPS OF MOVEMENTS

§ 1. Examples and definition of congruence groups of geometrical figures

1. Congruences of regular polygons in their planes

A large and very important class of groups, which comprises both finite and infinite groups, is formed by the "*congruence groups*" of geometrical figures. By a congruence of a given geometrical figure F we understand a movement of F (in space or in the plane) which transforms F into itself, that is to say which brings the figure F into coincidence with itself.

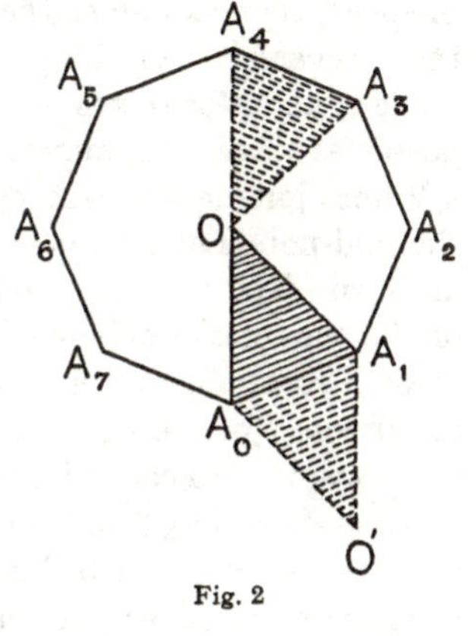

Fig. 2

We have already made ourselves familiar with some simple congruence groups, namely with the groups of rotations of regular polygons. Consider the regular polygon $A_0A_1 \ldots A_n$ in the plane (fig. 2), for example the regular octagon $A_0A_1A_2A_3A_4A_5A_6A_7$ (the vertices are all numbered in order, for example in the counter-clockwise sense).

We look for those movements of the polygon in its own plane which bring it into coincidence with itself. In these movements every vertex of the polygon must go over into a vertex, every side into a side, and the centroid O of the polygon must go over into itself. In a certain movement the vertex A_0 may for example go over into the vertex A_k (in the figure, $k = 4$).

Then the side A_0A_1 must either go over into the side A_kA_{k+1} or into the side A_kA_{k-1}. Suppose A_0A_1 went over into A_kA_{k-1}, then also the triangle A_0A_1O would go into triangle $A_kA_{k-1}O$. Now by a movement within its own plane, this triangle could be brought into the position of A_0A_1O', which is constructed by reflecting the triangle A_0A_1O in its side A_0A_1. This would prove that the triangle A_0A_1O could be transformed by a movement within its own plane into its reflection, and this is impossible.*

Therefore the side A_0A_1 must go over into the side A_kA_{k+1}. In the same way we convince ourselves that the side A_1A_2 must go over into $A_{k+1}A_{k+2}$, the side A_2A_3 into $A_{k+2}A_{k+3}$, and so on. In other words, the movement is a rotation of the polygon in its own plane and through an angle $k \cdot 2\pi/n$. Therefore

Every congruence of a regular n-gon in its own plane is a rotation of the polygon through an angle $k \cdot 2\pi/n$, where k is a whole number.

There are therefore n such congruences.

As we know, these rotations form a group.

2. Congruences of a regular polygon in three-dimensional space

We have carried through the foregoing investigations under the important assumption that only congruences of a polygon in its own plane were to be considered. If we investigate congruences of an n-gon in space, then as well as the above rotations there are also to be added the " reversals " of the polygon; these are the rotations through an angle π about the axes of symmetry of the polygon. A regular n-gon possesses n axes of symmetry. For n even, the axes of symmetry are the $n/2$ lines joining the pairs of opposite vertices and the $n/2$ lines joining the mid-points of the opposite sides. For n odd, the axes of symmetry are formed by the lines each of which joins a vertex to the mid-point of the side of the polygon lying opposite this vertex. The proof that these n rotations and n reversals of a regular n-gon constitute all the congruences of the n-gon, i.e. all the movements in space which bring the polygon into coincidence with itself, is essentially contained in the investigations in § 3 of this chapter. For the better understanding of this paragraph it may be left to the reader to turn back to it again later, in order to consider once more all the questions connected with the congruences of a regular polygon.

3. General definition of the congruence group of a given figure in space or in the plane

Suppose a figure F is given in space or in the plane. We consider the set of all congruences of this figure, i.e. all movements in space or in the plane which bring this figure into coincidence with itself.

* A rigorous proof of this impossibility, which is one of the basic facts in the geometry of the plane, would go beyond the scope of this book.

We define the sum $g_1 + g_2$ of two congruences g_1 and g_2 to be the movement which consists of the successive application of the movements g_1 and g_2 in this order. Evidently, under the assumption that g_1 and g_2 are congruences of the figure F, so also is $g_1 + g_2$. *The set of all congruences of the figure F forms a group with respect to the above definition of the operation of addition.* Indeed addition of movements satisfies the associative law. Furthermore, among the set of all congruences there is a null or " identical " congruence, namely that one which keeps every point of the figure fixed. Finally there corresponds to every congruence g the congruence $-g$ inverse to it (it sends every point back to its original position from the new position which it is occupying after the movement g).

§ 2. The congruence groups of a line, a circle, and a plane

The congruence groups of regular polygons are finite. We shall become acquainted with other finite congruence groups in this chapter, namely the congruence groups of certain polyhedra. But first of all we give some examples of infinite congruence groups.

The group of congruences of a straight line in any plane containing it forms the first example. This group consists of displacements of the line along itself (congruences of the first kind) and of rotations of the line through π radians in the selected plane and about any one of the points of the line (congruences of the second kind).

The congruence group of a line is non-commutative.—In order to convince ourselves of this it is sufficient to add together two congruences, one of the first kind and the other of the second kind. The result of this addition is changed if we change the order in which the terms are added.* Evidently we can obtain all the congruences of the second kind if we add to each possible displacement of the line an *arbitrary rotation* through π radians, that is to say a rotation through π radians about one particular, but arbitrarily chosen, point of the line.

The displacements of the line along itself form a subgroup of the group of all congruences of the line. These displacements are the only movements of the line inside itself. Now there corresponds in a unique way to each displacement of the line along itself a real number, which specifies the length and direction of the displacement. From this fact

* It is left to the reader to verify this by forming the sum of two arbitrary but definite congruences one of each kind, taking the terms in each of the two possible orders.

we easily conclude that the group of all displacements of a line along itself is isomorphic to the group of real numbers (where the group operation is ordinary arithmetical addition).

As a second example we consider the group of all congruences of a circle in its own plane. This group consists of all possible rotations of the circle in its plane about its centre, where as usual we regard as the identity any rotation which is an integral multiple* of 2π.

To every element of our group there corresponds in this way a definite angle. Let the radian measure of this angle be x. Now since angles differing from one another by an integral multiple of 2π define one and the same rotation of the circle, it follows that to each element of the rotation group of the circle there corresponds not a single number x but the set of all numbers of the form $x + 2\pi k$, where k is an arbitrary integer.

On the other hand, to every real number x there corresponds a unique rotation of the circle, namely the rotation through the angle whose radian measure is x. Therefore between the rotations of a circle and the real numbers we can set up the following correspondence:

To every real number x there corresponds a uniquely determined rotation, namely the rotation through the angle x. But conversely every rotation corresponds to, not just one, but infinitely many real numbers differing from each other by integral multiples of 2π.

The group of rotations of a circle is denoted by the Greek letter κ ("kappa"), from the word κύκλος (cyclos) which means a circle.

As a third example we choose to consider *the group of all movements of a plane* in itself. Moreover in this connection we consider not one but two planes, of which the first is fixed while the second one can be moved about, or more accurately can slide about, on the first one. We can picture the first fixed plane as a table with all its sides extended infinitely far, while the second sliding plane can be thought of as a sheet of glass lying on the table, and likewise with all its sides infinitely extended. We are therefore thinking of the totality of possible movements of the sheet which keep it always in contact with the table.†

In the group of all movements of a plane in itself there are infinitely many subgroups. Of these we mention first the infinitely many rotation groups: The set of all rotations of the plane about an arbitrary,

* If the meaning of the following investigations is not readily intelligible to the reader he can pass straight on to the next example and return to this one after reading Chapter VIII.

† In particular therefore a rotation of the sheet about an axis lying wholly on the table is not allowed.

but definite, point of the plane forms a group, and we easily recognize that this group is isomorphic to the group κ. It follows in particular that each of these groups is commutative. In the group of all movements of the plane in itself, as well as the rotation groups, there are the subgroups of parallel displacements along different lines: Given a line g we can displace the plane along this line, and the line g and every line parallel to it will evidently be transformed into itself. There are two mutually opposite directions possible for these displacements along the line g. The set of all such displacements forms a group, *the group of displacements or parallel-displacements of the plane along a given line*; it is evidently a subgroup of the group of all movements of the plane in itself.

Every displacement along a line g is characterized by the magnitude and direction of a certain vector v which lies along the line g and issues from a selected (fixed once and for all) point O of this line (fig. 3). In our displacement the point O moves to the end point of the vector v. It follows from all this that *the group of all displacements of the plane along a given line g is isomorphic to the group of all real numbers* (with ordinary addition as the group operation).

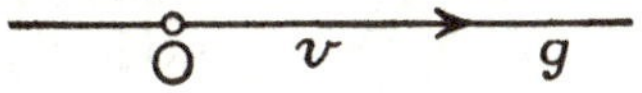

Fig. 3

We consider two displacements v and v' of the plane along two non-parallel lines * g and g' (fig. 4).

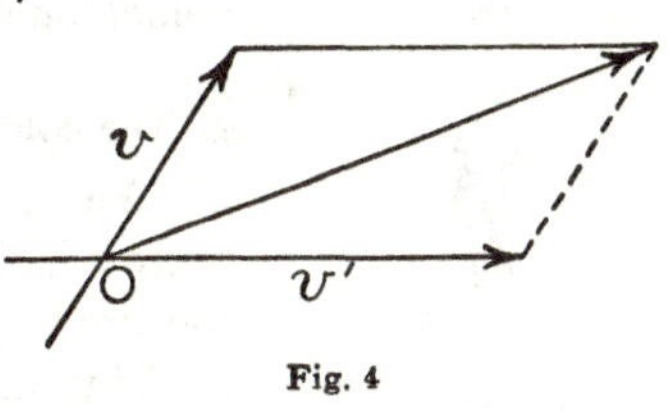

Fig. 4

The resultant of these two displacements is equivalent to the displacement of the plane along the diagonal of the parallelogram formed by v and v' (" parallelogram rule " for addition of vectors).

Therefore the sum of any two displacements of the plane is again a displacement of the plane. It does not depend on the order of the terms in the sum. From this it follows that the set of all displacements of the plane along all possible lines is a commutative subgroup of the group of movements of the plane in itself.

The last two groups which we have just considered, namely the groups of movements of a circle and of a plane, have the following in common: These groups consist of movements of the corresponding forms *inside themselves*. In other words, throughout each movement

* Two displacements along two parallel lines are obviously equivalent to displacements along *one* line (namely along either of the two given lines or just as well along any third line which is parallel to the given ones).

the form considered, the circle or the plane, remains in coincidence with itself. This property is no longer possessed by the congruences of regular polygons. Indeed for them the final and initial positions of the moving figure coincide, but the intermediate positions which the figure takes in the course of its movement differ from its initial and final positions. The same is true also of the movements of polyhedra to which we turn directly.

§ 3. The rotation groups of a regular pyramid and of a double pyramid

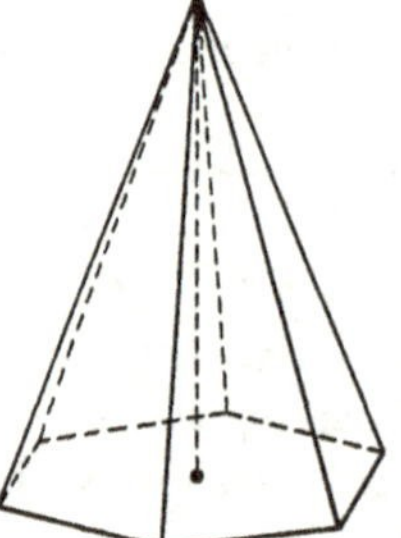

Fig. 5

1. The pyramid

The group of rotations about its axis of a regular pyramid whose base is an n-gon (fig. 5) is evidently isomorphic to the group of rotations in its plane of a regular n-gon. This group is therefore cyclic of order n. We easily convince ourselves that the rotations of the pyramid about its axis [through angles of $0, 2\pi/n, \ldots, (n-1)2\pi/n$] are the only movements which bring the pyramid into coincidence with itself (at any rate when $n > 3$).

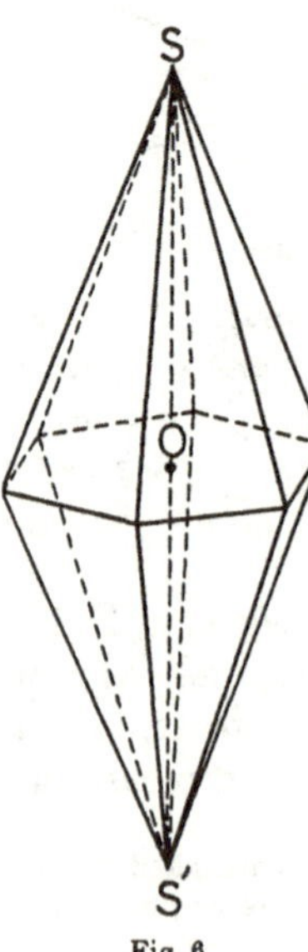

Fig. 6

2. The double pyramid

We now define the congruence group of a body which is called a "regular n-pointed double pyramid" (fig. 6).

This body consists of a regular pyramid whose base is an n-gon and its mirror image in its base. We shall prove that the congruence group of the double pyramid consists of the following elements:

1. The rotations about the axis of the pyramid (through angles of $0, 2\pi/n, \ldots, (n-1)2\pi/n$).

2. The so-called reversals, that is to say the rotations through π about each of the axes of symmetry of the base of the double pyramid, i.e. of the regular polygon which both the pyramids forming the double pyramid have as their common base. As we have seen there are n such axes of symmetry and therefore n reversals.

The number of all these movements is therefore equal to $2n$. In order to convince ourselves that, except in the case $n = 4$, there are no other movements which bring the n-pointed double pyramid into coincidence with itself, we first observe that for $n \neq 4$ every congruence of the double pyramid must either keep the points S and S′ fixed (*congruences of the first kind*) or must interchange them (*congruences of the second kind*). Further, in any congruence the base of the double pyramid must go over into itself. Finally we remark that the addition of (that is to say, the successive application of) two congruences of the first kind is again a congruence of the first kind, the addition of a congruence of the first kind to a congruence of the second kind is a congruence of the second kind, and the addition of two congruences of the second kind is a congruence of the first kind.

In this connection, the sum of two congruences, of which one is of the first kind and the other of the second kind, depends on the order in which the terms are added. Indeed if a is a congruence of the first kind and b a congruence of the second kind then $a + b = b - a$.

We consider first of all congruences of the first kind. These congruences do not overturn the base, that is to say the base does not move out of its own plane. It can therefore only be subject to a rotation through one of the angles

$$0,\ 2\pi/n,\ \ldots,\ (n-1)2\pi/n$$

Each of these movements is therefore a rotation through one of these angles, about the axis of the double pyramid.

There are therefore exactly n congruences of the first kind (including the identity). These congruences are none other than the rotations of the double pyramid through angles $0,\ 2\pi/n,\ \ldots,$ $(n-1)2\pi/n$ about its axis.

Suppose given an arbitrary but fixed congruence of the second kind, that is to say a congruence of the double pyramid which interchanges the points S and S′.

If we apply, after this congruence of the second kind, an arbitrary but fixed reversal of the double pyramid, that is to say, a movement which consists of a rotation of the double pyramid through the angle π about *an arbitrarily chosen* axis of symmetry of the base, then we obtain a congruence of the first kind,* and therefore a rotation of the double pyramid about its axis.

Hence every congruence of the second kind added to a fixed reversal

* Indeed every one of these reversals is a congruence of the second kind, and the sum of two congruences of the second kind is a congruence of the first kind.

results in a congruence of the first kind. From this it follows that: Every congruence of the second kind consists of a suitable congruence of the first kind following or followed by an arbitrarily chosen but fixed reversal. It follows further that the number of congruences of the first kind is equal to the number of congruences of the second kind, and therefore evidently this number is equal to n.

On the other hand it is clear that all reversals are congruences of the second kind. Since there are exactly n such reversals they evidently comprise all the congruences of the second kind.

Hence for $n \neq 4$ we have established the following: *The congruence group of an n-pointed double pyramid is a non-commutative group of order* $2n$,* *which consists of n rotations about the axis* SS′ *of the double pyramid and of n reversals, that is to say, rotations through an angle* π *about the axes of symmetry of the base of the double pyramid. We obtain all the n reversals by adding a single one of them to the n rotations of the double pyramid about its axis* SS′.

Further, since we obtain all the rotations of the double pyramid about its axis by repeated addition of a single one of these rotations, namely the rotation through the angle $2\pi/n$, therefore the group of all congruences possesses a system of generators which consists of two elements: of the rotation through the angle $2\pi/n$ and of an arbitrary reversal.

The case $n = 4$ is an exception, in that a special case of a four-pointed double pyramid is an octahedron, and this possesses, as we shall see below, not 8 but 24 congruences. This is because in the regular octahedron we can interchange the vertex S not only with the vertex S′ but also with each of the vertices of the base. One of the necessary conditions for this to be allowable, namely that to each vertex there corresponds the same number of faces and edges, is evidently already satisfied for an arbitrary 4-pointed double pyramid. Moreover, for a regular octahedron, the angles in the faces and also the angles between the faces corresponding to any two arbitrary vertices are equal, and therefore the adjacent faces and edges are congruent.

3. Degenerate cases: The rotation groups of an interval and of a rhombus

The smallest number of vertices which a polygon can have is three. However, as is well known, we can regard an interval of a straight line as a " degenerate " polygon or as a " polygon with 2 vertices ". This is also justified in particular by the fact that the congruence group of

* It is called the *dihedral group* of order $2n$.

an interval in any plane containing it is a cyclic group of order 2. It apparently consists of the identical congruence and the rotation of the interval through an angle π about its mid-point.

Similarly an isosceles triangle can be regarded as a degenerate case of a regular pyramid: The congruence group of an isosceles triangle in space is a group of order 2.

Furthermore a degenerate double pyramid is evidently a rhombus. The group of congruences or rotations of a rhombus in space consists of four elements: The identical transformation a_0, the rotations a_1 and a_2 through an angle π about the diagonals, and the rotation a_3 in its plane through an angle π about its centroid; this is the sum of the rotations a_1 and a_2.* The addition table of our group takes the following form:

	a_0	a_1	a_2	a_3
a_0	a_0	a_1	a_2	a_3
a_1	a_1	a_0	a_3	a_2
a_2	a_2	a_3	a_0	a_1
a_3	a_3	a_2	a_1	a_0

which is therefore identical with the addition table of Klein's four-group, considered as the second example in Chapter I, § 1, section 3. We easily convince ourselves of this directly, or alternatively by considering instead of the rotation group of the rhombus the isomorphic group of permutations of its four vertices A, B, C, D. Evidently the rotations a_0, a_1, a_2, a_3 correspond to the following permutations of the vertices:†

$$\begin{pmatrix} A\,B\,C\,D \\ A\,B\,C\,D \end{pmatrix} \quad \begin{pmatrix} A\,B\,C\,D \\ B\,A\,C\,D \end{pmatrix} \quad \begin{pmatrix} A\,B\,C\,D \\ A\,B\,D\,C \end{pmatrix} \quad \begin{pmatrix} A\,B\,C\,D \\ B\,A\,D\,C \end{pmatrix}$$

* If we regard one of the diagonals of the rhombus as the " baseline " and the other as the axis of the corresponding degenerate double pyramid, then we may interpret these four congruences as rotations about the axis and reversals, as in the non-degenerate case.

† We denote by a_1 the rotation about the diagonal CD, and by a_2 that about the diagonal AB.

§ 4. The rotation group of a tetrahedron *

In order to determine all the congruences of the tetrahedron $A_0A_1A_2A_3$ (fig. 7) we first of all consider those which keep one particular vertex fixed, say A_0. These congruences then carry the triangle $A_1A_2A_3$ into itself by rotating it about its centroid B_0 through the angles 0, $2\pi/3$, $4\pi/3$. From this it follows that there are exactly three congruences of the tetrahedron $A_0A_1A_2A_3$ which keep the vertex A_0 fixed: the identical congruence a_0 which keeps every element of the tetrahedron fixed, and the two rotations a_1 and a_2 through the angles $2\pi/3$ and $4\pi/3$ respectively about the axis A_0B_0. We now denote by x_i any particular congruence of the tetrahedron which carries the vertex A_0 into the vertex A_i $(i = 1, 2, 3)$.† By x_0 we denote again the identical congruence.

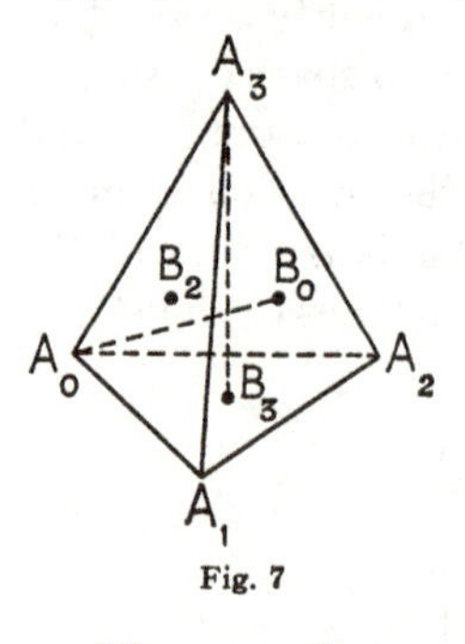

Fig. 7

We prove that every congruence b of the tetrahedron can be written in the form

$$b = a_i + x_k \tag{1}$$

where $i = 0, 1, 2$ and $k = 0, 1, 2, 3$ are uniquely determined by b. (This last assertion means that if $b = a_i + x_k$, $b' = a_i' + x_k'$, and if at least one of the inequalities $i \neq i'$, $k \neq k'$ is true, then $b \neq b'$.)

Let us therefore suppose given any congruence b. It carries over the vertex A_0 into a certain uniquely determined vertex A_k, where $k = 0, 1, 2, 3$. But then the congruence $b - x_k$ leaves the vertex A_0 fixed and it is therefore equal to a uniquely determined a_i, so that we have $b - x_k = a_i$, that is $b = a_i + x_k$; here i and k are uniquely determined. Since also conversely to every pair (i, k) there corresponds by (1) a particular congruence of the tetrahedron, there is a one-to-one correspondence between the set of all congruences of the tetrahedron and all pairs (i, k) where i takes the values 0, 1, 2 and k the values 0, 1, 2, 3. It follows from this that there are exactly 12 congruences of the tetrahedron. Now every congruence of the tetrahedron gives rise to a certain permutation of the vertices, and therefore to a certain permutation of the corresponding numbers 0, 1, 2, 3. But now there are 24

* By a tetrahedron we understand here and always in what follows a *regular* tetrahedron.

† The vertex A_0 can be mapped into A_1 and A_3, for example by rotations about the axis A_2B_2 (which joins A_2 to the centroid of the opposite face). A_0 goes over into A_2, for example by a rotation about the axis A_3B_3.

permutations of four elements, and of these, as we have just seen, only 12 can be associated with movements of the tetrahedron in space. We will investigate which of these movements correspond to which permutations.

For the sake of brevity we refer to any line joining a vertex A_i of the tetrahedron to the centroid of the face lying opposite A_i as a *median.* We use the word *edge-bisector* to describe any line which joins the midpoints of two opposite edges of the tetrahedron.

To every median there correspond two congruences of the tetrahedron other than the identity, namely the rotations through angles of $2\pi/3$ and $4\pi/3$ about the median. Altogether therefore we obtain eight rotations which we can represent in the following way as permutations of the suffixes of the vertices:

$$\begin{array}{llll} a_1 = \begin{pmatrix} 0123 \\ 0231 \end{pmatrix} & a_2 = \begin{pmatrix} 0123 \\ 0312 \end{pmatrix} & a_3 = \begin{pmatrix} 0123 \\ 2130 \end{pmatrix} & a_4 = \begin{pmatrix} 0123 \\ 3102 \end{pmatrix} \\ a_5 = \begin{pmatrix} 0123 \\ 1320 \end{pmatrix} & a_6 = \begin{pmatrix} 0123 \\ 3021 \end{pmatrix} & a_7 = \begin{pmatrix} 0123 \\ 1203 \end{pmatrix} & a_8 = \begin{pmatrix} 0123 \\ 2013 \end{pmatrix} \end{array} \tag{2}$$

About each edge-bisector there is a rotation through the angle π, which is different from the identity, and since there are three edge-bisectors this results in another three rotations; they can be written as permutations in the following way:

$$a_9 = \begin{pmatrix} 0123 \\ 1032 \end{pmatrix} \quad a_{10} = \begin{pmatrix} 0123 \\ 2301 \end{pmatrix} \quad a_{11} = \begin{pmatrix} 0123 \\ 3210 \end{pmatrix} \tag{3}$$

These eleven rotations together with the identical congruence (" identical rotation ") yield just the twelve congruences of the tetrahedron. Every one of them is a rotation about one of the seven axes of symmetry* of the tetrahedron. Therefore the group of these congruences is also called *the rotation group of the tetrahedron.*

We easily verify that all the permutations (2) and (3) are even. But since there are altogether twelve even permutations on four elements, in this case the vertices of the tetrahedron, we evidently

* These seven axes of symmetry consist of the four medians and the three edge-bisectors of the tetrahedron. More generally we mean by an axis of symmetry of a geometrical figure any line about which the figure can be rotated through an angle different from zero, so as to be carried into coincidence with itself. In this connection we remark that every movement of a rigid body in space leaving a certain point O fixed is a rotation of this rigid body about a certain axis passing through the point O.

have *a one-to-one and indeed an isomorphic correspondence between the rotation group of the tetrahedron and the alternating group of permutations on four elements.*

We wish now to investigate what subgroups are possessed by the rotation group of the tetrahedron.

In this rotation group, as in every group, there are the two *improper* subgroups: firstly the whole group itself, and secondly the subgroup which consists only of the null element. Let us concern ourselves with the remaining subgroups, the *proper* subgroups of the rotation group of the tetrahedron. There are exactly eight of these.

First of all we remark that the sum of the rotations through the angle π about two different edge-bisectors is a rotation through π about the third edge-bisector; this can be verified geometrically, but also by adding together any two of the permutations (3). From this it follows that the rotations through the angle π about the three edge-bisectors together with the identical rotation form a group of order four. It is isomorphic to Klein's four-group and therefore also to the group of rotations of the rhombus. We denote this group by H. Among all the subgroups of the rotation group of the tetrahedron this one has the highest order. It contains three subgroups of the second order, each of which consists of rotations through angles of 0 and π about one of the given edge-bisectors. We denote these subgroups by H_{01}, H_{02}, H_{03}. Besides those named there are four more subgroups which are of order 3, namely the groups

$$H_i \; (i = 0, 1, 2, 3)$$

each of which consists of three rotations through angles of 0, $2\pi/3$, $4\pi/3$ about one of the medians.

In order to prove that in the rotation group of the tetrahedron there are no other subgroups, it is sufficient to show that any two elements different from the null element, which are taken either from two different groups H_i, or one of which is taken from a group H_i and the other from a group H_{0k}, form a system of generators for the whole rotation group of the tetrahedron. To achieve this it is again sufficient to consider any two of the elements a_1, a_3, a_5, a_7, say a_1 and a_3, or one of the elements a_1, a_3, a_5, a_7 and one of the elements a_9, a_{10}, a_{11}. We leave it to the reader to carry through the geometrical proof, and therefore to show that every rotation of the tetrahedron can be generated by any one of the pairs of rotations mentioned. We can establish the same result also by calculation. The following identities show that for example the elements a_1 and a_3 form a system of generators of the rotation group of the tetrahedron:

$$
\begin{aligned}
a_0 &= a_1 - a_1 & a_7 &= a_1 + a_3 - a_1 \\
a_2 &= 2a_1 & a_8 &= 2a_1 + a_3 \\
a_4 &= 2a_3 & a_9 &= -a_3 + a_1 + 2a_3 \\
a_5 &= -a_3 + a_1 + a_3 & a_{10} &= a_1 + a_3 \\
a_6 &= -a_3 + 2a_1 + a_3 & a_{11} &= a_3 + a_1
\end{aligned}
$$

It must not be supposed that every element can be represented in terms of the generators in a *unique* way. Thus we have

$$a_7 = a_1 + a_3 - a_1$$

and at the same time

$$a_7 = -a_3 - a_1 + a_3 + a_1 + a_3$$

The rotation group of the tetrahedron is non-commutative: thus

$$a_1 + a_3 = a_{10}$$

while

$$a_3 + a_1 = a_{11}$$

§ 5. The rotation group of a cube and of an octahedron *

In order to obtain all the congruences of a cube we proceed just as for the tetrahedron. We consider first of all only those congruences of the cube ABCDA'B'C'D' (fig. 8) which carry one vertex, say A, over into itself.

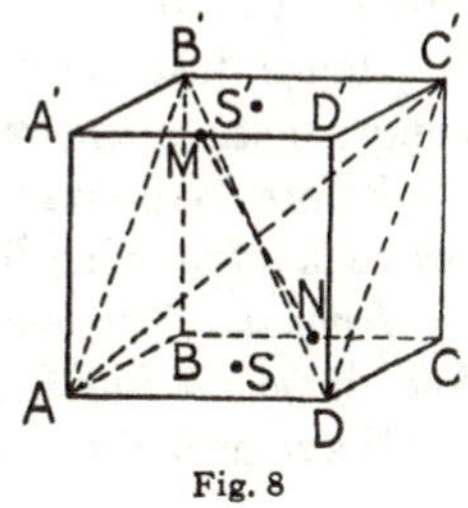

Fig. 8

In each congruence of a cube, vertices go over into vertices, edges into edges, and faces into faces; also the diagonals of the cube go over into each other. A given congruence which

* Just as in the case of the tetrahedron we always understand by the word " octahedron " a *regular* octahedron.

leaves the vertex A fixed leaves also the diagonal AC′ fixed since there exists only one diagonal of the cube passing through A. Therefore this congruence is a rotation of the cube about the diagonal AC′. There are just two such rotations different from the identity, namely those through angles of $2\pi/3$ and $4\pi/3$.

There are therefore altogether three congruences of the cube which carry the vertex A over into itself. We can find the corresponding rotations for all eight vertices of the cube exactly as for the vertex A. If we argue as in the case of the tetrahedron then we easily deduce that altogether there are $3 \times 8 = 24$ congruences of the cube.

We will investigate these congruences more closely. First of all we remark that a cube has 13 axes of symmetry: the four diagonals of the body, the three lines, each of which joins the centroids of a pair of opposite faces, and the six lines each of which joins the mid-points of a pair of opposite edges of the cube. About each of the four diagonals there are two rotations of the cube different from the identity, which carry the cube over into itself. Altogether therefore there are eight rotations about the diagonals.

About each of the lines joining the centroids of a pair of opposite faces there are three rotations different from the identity, and hence altogether nine such rotations.

Finally there is a rotation different from the identity through π about each line joining the mid-points of a pair of opposite edges, and therefore altogether six such rotations.

Thus we have $8 + 9 + 6 = 23$ rotations, different from the identity, carrying the cube over into itself. If to them we add further the identical rotation then we obtain 24 congruences, and therefore all the possible congruences of the cube.

Therefore *all the congruences of the cube consist of rotations about its axes of symmetry*.

Hence, just as for the tetrahedron, we usually speak of the group of congruences of the cube as the rotation group of the cube.

Before we go on to consider the structure of the rotation group we prove the following lemma:

*Lemma.—The only rotation of the cube which carries over each of the four diagonals into itself is the identical rotation.**

We remark first that any rotation which carries over into themselves

* The following remark should be noted carefully: If a given diagonal, say AC′, goes over into itself by a given rotation then this does not imply that the vertices joined by this diagonal (in our case the vertices A and C′) necessarily remain fixed. They may be interchanged, that is to say, A may go into C′ and C′ into A.

any two diagonals of the cube, say AC′ and DB′, also carries over into itself the diagonal plane ADC′B′ (fig. 8). Now every rotation other than the identity which carries a certain plane over into itself has as axis either a line lying in this plane—in which case the angle of rotation is equal to π—or a line perpendicular to this plane. But now a rotation of the plane through an angle π about an axis lying in the plane carries over into themselves only those lines of the plane which are perpendicular to the axis, except for the axis itself. Since the quadrilateral ADC′B′ is not a square its diagonals, since they are not at right angles to each other, cannot both go over into themselves by any rotation of the cube about an axis lying in the plane of the quadrilateral. Therefore AC′ and DB′ can only go over into themselves by a rotation of the cube about an axis perpendicular to the plane ADC′B′. This axis is the line MN joining the mid-points of the sides A′D′ and BC. The only rotation of the cube about the line MN, different from the identity, is the rotation through the angle π. Therefore this is the only rotation which carries over into itself each of the diagonals AC′ and DB′. However, the other two diagonals BD′ and CA′ are interchanged by this rotation, and so there exists no rotation other than the identity which carries over into themselves all the four diagonals.

Therefore every rotation of the cube different from the identity subjects the four diagonals to a non-identical permutation. From this it follows that in two different rotations a and b the diagonals also undergo different permutations. Indeed if two rotations a and b resulted in the same permutation of the diagonals, then in the rotation $a - b$ all the diagonals would remain fixed, and therefore $a - b$ would be the identical rotation, so that a and b would coincide.

Thus there correspond to the 24 distinct rotations of the cube distinct permutations of the diagonals which are produced by these rotations. But it is well known that there are $1 . 2 . 3 . 4 = 24$ permutations of four elements.

From this it follows that there exists a one-to-one correspondence between the group of all rotations of the cube and the group of all permutations of its four diagonals. Since in our correspondence addition of rotations corresponds exactly to addition of permutations,* we have the following theorem:

* By this we mean that to the sum of two rotations there corresponds the sum of the permutations corresponding to these rotations, where the elements in each case are added in the same order. Merely a one-to-one correspondence between the rotations and the permutations of the diagonals would on the other hand imply only an otherwise completely arbitrary association of the rotations with the permutations.

The rotation group of a cube is isomorphic to the group of all permutations on four elements.

Among the subgroups of the rotation group of the cube, we mention first of all those cyclic subgroups of orders two, three, and four, each of which consists of rotations about one of the thirteen axes of symmetry of the cube. There are six cyclic subgroups of order two corresponding to the number of axes joining the mid-points of opposite edges; four cyclic subgroups of order three, equal to the number of diagonals; three cyclic subgroups of order four corresponding to the number of lines joining the centroids of opposite faces.

The following subgroups which also occur are very much more interesting:

(*a*) The subgroup of order twelve which consists of those rotations which carry over into themselves both of the tetrahedra ACB′D′ and BDA′C′ (fig. 9) inscribed in the cube. This subgroup consists of the 2×4 rotations about the diagonals different from the identity, the three rotations through an angle π about the axes joining the centroids of opposite faces, and the identical rotation.

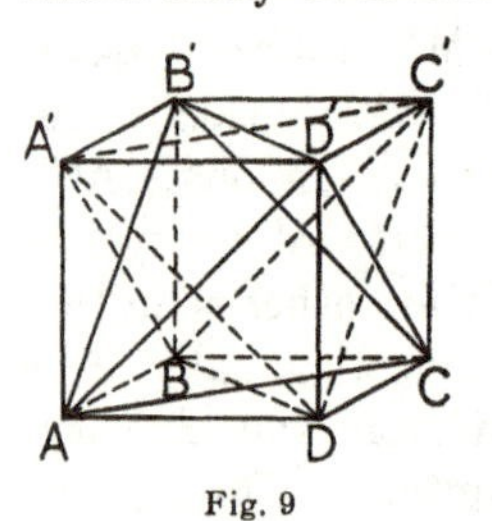

Fig. 9

(*b*) Three subgroups of order eight which are isomorphic to the group of the square double pyramid. Each of these subgroups consists of those rotations of the cube which carry over into itself one of the lines joining the centroids of two opposite faces, for example the points S and S′. (The octahedron inscribed in the cube is a special case of the square double pyramid. The group of those of its rotations leaving fixed two of its vertices S and S′ or interchanging them is evidently identical with the group of the square double pyramid.)

Such a subgroup of order eight consists of the following eight rotations: Four rotations about the axis SS′ (including the identity); two rotations through the angle π about the axes joining respectively the mid-points of the edges AA′ and CC′, and BB′ and DD′; and two rotations through the angle π about the axes joining respectively the centroids of the faces ABB′A′ and CDD′C′, and ADD′A′ and BCC′B′.

(*c*) A subgroup of order four which consists of the identical rotation and of three rotations through the angle π about the axes joining the centroids of two opposite faces. This group consists of those rotations which occur in each of the three subgroups of order eight mentioned above. It is commutative and isomorphic to the rotation group of the rhombus, and therefore also to Klein's four-group.

The group of congruences or rotations of a regular octahedron is isomorphic to the rotation group of a cube.

In order to convince ourselves of this it is sufficient to describe round the regular octahedron a cube (fig. 10) or equally well to inscribe in the regular octahedron a cube (fig. 11). To each congruence of the octahedron there corresponds a certain congruence of the cube, and conversely.

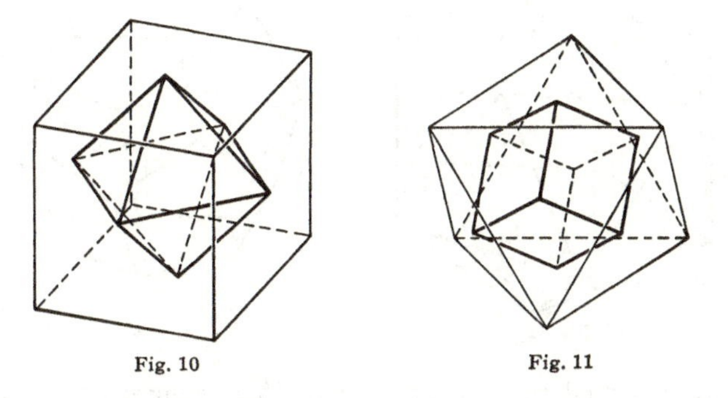

Fig. 10 Fig. 11

At this point there enters in the idea of a dual relationship existing between a cube and an octahedron; we will go into this matter more closely now.

First of all we say that two elements (vertices, edges, faces) of any polyhedron are *associated* if one of these two elements is a constituent part of the other. Hence a vertex and a face containing this vertex as one of its vertices, a face and an edge of this face, a vertex and an edge having this vertex as an endpoint are pairs of associated elements.

Two polyhedra are called *dual* if the elements of the one polyhedron can be put in one-to-one correspondence with the elements of the other in such a way that pairs of associated elements of the one polyhedron correspond to pairs of associated elements of the other, and further that

to the vertices of the first polyhedron there correspond the faces of the second,

to the edges of the first polyhedron there correspond the edges of the second,

to the faces of the first polyhedron there correspond the vertices of the second.

We easily see that the cube and the octahedron are dual to each other in this sense. The tetrahedron is self-dual.

§ 6. The rotation group of an icosahedron and of a dodecahedron.* General remarks about rotation groups of regular polyhedra

There remain two of the five regular polyhedra still to be investigated: the icosahedron and the dodecahedron (figs. 12 and 13). These

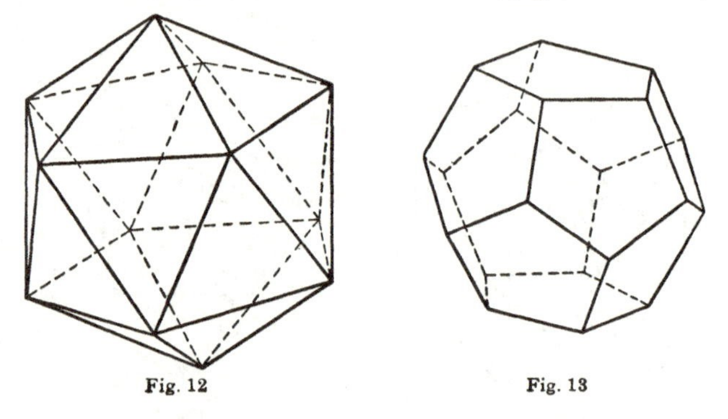

Fig. 12 Fig. 13

polyhedra are dual to each other, and their congruence groups are isomorphic.

In order to convince ourselves of this it is sufficient to inscribe the icosahedron in the dodecahedron (fig. 14) or to inscribe the dodecahedron in the icosahedron (fig. 15). Thus we need only make ourselves familiar with the congruence group of the icosahedron. In order to

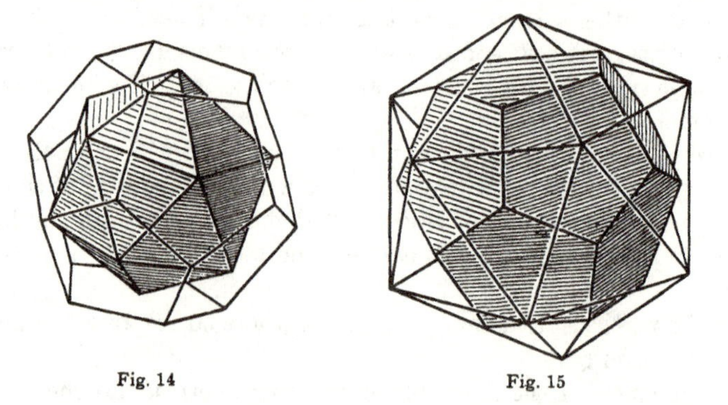

Fig. 14 Fig. 15

determine the number of its elements we proceed just as for the tetrahedron and the cube. We consider first of all those congruences of the icosahedron which leave a certain one of its vertices fixed.

* We mean again a *regular* icosahedron and a *regular* dodecahedron.

There are five such congruences, namely five rotations about the axis joining this vertex to the one lying opposite to it. Since there are twelve vertices (therefore $k = 0, 1, \ldots, 11$) the number of congruences of the icosahedron is equal to $5 \times 12 = 60$.

All these congruences are rotations of the icosahedron about its axes of symmetry. In particular the icosahedron possesses the following axes of symmetry:

Six axes joining opposite vertices; about each of these there are four rotations different from the identity (through angles of $2\pi/5$, $4\pi/5$, $6\pi/5$, $8\pi/5$) which bring the icosahedron into coincidence with itself; therefore altogether we obtain $4 \times 6 = 24$ rotations.

Ten axes joining the centroids of opposite faces; about each of these axes there are two rotations different from the identity (through angles of $2\pi/3$ and $4\pi/3$), and therefore altogether 20 rotations.

Fifteen axes joining the mid-points of opposite edges, and about each of these a single rotation other than the identity through an angle π.

Therefore there are $24 + 20 + 15$ rotations, and taking into account the identity, therefore altogether 60 rotations.

It follows from this that the icosahedron possesses *exactly* 31 axes of symmetry.

Since the rotation group of the icosahedron is particularly complicated we shall not investigate it any further here. We simply mention that it is isomorphic to the alternating group of permutations on five elements.

The rotation groups of the regular polygons and polyhedra were defined as their groups of congruences.

We consider now as it were two spaces, one of which is embedded in the other. We picture one space as a rigid body, all of whose sides are indefinitely extended, and we call it the *rigid* space, while the other one is thought of as *empty* space.

We imagine the rigid space to be embedded in, and able to move about in, the empty space. Our polyhedron appears as a fixed part of the rigid space, and it can therefore only move with it.* With this interpretation we can consider all rotations of the " rigid " space about any axis in the " empty " space which bring the given polyhedron into coincidence with itself, that is to say which carry it over into itself. Since every congruence of the polyhedron considered turns out to be a

* An example of a " rigid " space moving about in an underlying " empty " space is provided by a glass sheet moving on a tabletop (see § 2, third example).

rotation about a suitable axis, and since every rotation about an axis can be regarded as defining a rotation of the whole space about this axis, it follows that the group of congruences of a given polyhedron is isomorphic to the group of rotations of space which carry this polyhedron over into itself. We mean precisely this group usually when we speak of the rotation group of a regular polyhedron. Often we speak of it simply as " the group of the regular polyhedron ".

The groups of the regular pyramids (that is to say the finite cyclic groups), the groups of the double pyramids (the *dihedral groups*), and the groups of the regular polyhedra just considered are the only *finite* subgroups of the group of all rigid movements in space.

EXERCISES ON CHAPTER V

1. Make an addition table for the group of congruences in space of an equilateral triangle.

2. Prove that the group of all displacements of the plane is isomorphic to the group of complex numbers, with ordinary addition as the group operation.

3. Prove that the set of all rotations of the plane in itself (about all possible points of the plane) does not form a group.

4. Make an addition table for the group of rotations of a regular tetrahedron.

5. Prove that a set E of elements of a group G is a system of generators of this group if and only if no proper subgroup of G exists which contains all the elements of the set E.

6. Make use of the result proved in Ex. 5 to find all possible systems of generators of the rotation group of the tetrahedron which consist of at most three elements. (We see from this example that it is possible for a finite group to possess many different systems of generators.)

Chapter VI

INVARIANT SUBGROUPS

§ 1. Conjugate elements and subgroups

1. Transformation of one group element by another

We consider in the group G two arbitrary elements a and b. The element

$$-b + a + b$$

is called the *transform* of the element a by b.

We wish to investigate under what conditions the equation

$$-b + a + b = a \tag{1}$$

is valid. If equation (1) is true, on adding b to the left of both sides, we obtain

$$a + b = b + a \tag{1'}$$

Therefore if (1) is true so also is (1′), i.e. the elements a and b are commutable. Conversely if (1′) is true then also

$$-b + a + b = -b + b + a = a$$

and hence (1) is true. Therefore:

For the validity of equation (1) *for given a and b*, i.e. for an element a to be equal to its transform by b, *it is necessary and sufficient that a and b shall be commutable*, i.e. that equation (1′) shall be valid.

In particular equation (1) is true for any two elements a and b of a commutative group.

In order to illustrate the concept of the transform we consider the group G of all permutations on n elements. Let

$$a = \begin{pmatrix} 1 & 2 & 3 & \dots & n \\ a_1 & a_2 & a_3 & \dots & a_n \end{pmatrix} \qquad b = \begin{pmatrix} 1 & 2 & 3 & \dots & n \\ b_1 & b_2 & b_3 & \dots & b_n \end{pmatrix}$$

Then evidently

$$-b = \begin{pmatrix} b_1 & b_2 & b_3 & \dots & b_n \\ 1 & 2 & 3 & \dots & n \end{pmatrix}$$

$$-b + a = \begin{pmatrix} b_1 & b_2 & b_3 & \dots & b_n \\ a_1 & a_2 & a_3 & \dots & a_n \end{pmatrix} \tag{2}$$

$$-b + a + b = \begin{pmatrix} b_1 & b_2 & b_3 & \dots & b_n \\ b_{a_1} & b_{a_2} & b_{a_3} & \dots & b_{a_n} \end{pmatrix}$$

The formula (2) can be expressed in the form of the following rule: Let

$$a = \begin{pmatrix} 1 & 2 & 3 & \dots & n \\ a_1 & a_2 & a_3 & \dots & a_n \end{pmatrix} \quad \text{and} \quad b = \begin{pmatrix} 1 & 2 & 3 & \dots & n \\ b_1 & b_2 & b_3 & \dots & b_n \end{pmatrix}$$

In order to obtain the transform of the permutation a by the permutation b we must apply the permutation b to both rows of the permutation a when it is written in the usual form.

We will illustrate this rule further by an example. Suppose for example that $n = 3$ and

$$a = \begin{pmatrix} 1 & 2 & 3 \\ 2 & 1 & 3 \end{pmatrix} \qquad b = \begin{pmatrix} 1 & 2 & 3 \\ 3 & 2 & 1 \end{pmatrix}$$

We obtain

$$-b + a + b = \begin{pmatrix} 3 & 2 & 1 \\ 2 & 3 & 1 \end{pmatrix} = \begin{pmatrix} 1 & 2 & 3 \\ 1 & 3 & 2 \end{pmatrix} \neq a$$

The rule just introduced is better understood if we make use of the concept of a mapping or function.*

The permutation a specifies a function $y = f(x)$

$$(x = 1, 2, 3, \dots, n; \quad y = 1, 2, 3, \dots, n)$$

in which to two different values of x there always correspond two different values of y. The permutation b is a function $y = \varphi(x)$ of the same kind as $f(x)$. The permutation $-b + a + b$ is then the function $y = F(x)$ defined by the formula

$$F(x) = \varphi\{f[\varphi^{-1}(x)]\} \tag{3}$$

We obtain this function by associating with the element $\varphi(x)$ the element $\varphi[f(x)]$. This is immediately obvious if we replace x by $\varphi(x)$ in formula (3) and observe that

$$\varphi^{-1}[\varphi(x)] = x$$

* See Appendix § 4.

As x runs through all the numbers 1, 2, 3, ..., n so also does $\varphi(x)$ only in a different order. The function $F(x)$, and therefore the permutation $-b + a + b$, is uniquely defined by means of the formula

$$F[\varphi(x)] = \varphi[f(x)] \tag{4}$$

The formula (4) is only another way of writing (2). Finally if we denote $f(x)$ by y then we can formulate the result stated above in the following way:

By the permutation $F(x)$ the element $\varphi(x)$ is replaced by the element $\varphi(y)$.

Since every finite group is isomorphic to a certain group of permutations the formula (2) illustrates the concept of a " transform " at least for all finite groups.

2. Transformation of elements in the group of the tetrahedron

As a further example we consider the rotation group of the tetrahedron ABCD (fig. 16).

Let a be the rotation of the tetrahedron through the angle π about the axis MN joining the mid-points of the sides BC and AD; let b be the rotation about the axis DO which carries C into A, A into B, B into C, i.e. which replaces A by C, B by A, C by B. Then $-b + a + b$ is the rotation through the angle π about the axis PQ joining the mid-points of the sides AB and CD. We can convince ourselves of this directly, or alternatively by interpreting the rotation a as the permutation $\begin{pmatrix} A & B & C & D \\ D & C & B & A \end{pmatrix}$ of the vertices and the rotation b as the permutation $\begin{pmatrix} A & B & C & D \\ C & A & B & D \end{pmatrix}$.

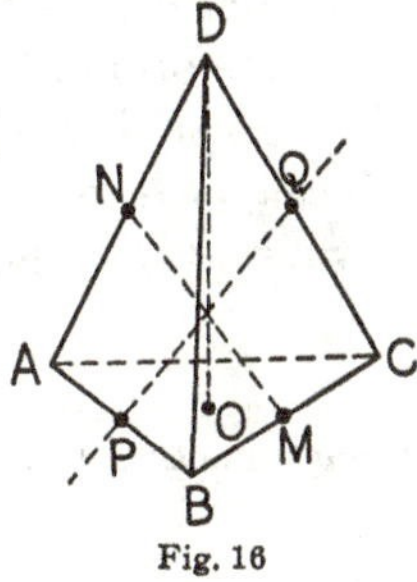

Fig. 16

If we now subject each row in the expression $\begin{pmatrix} A & B & C & D \\ D & C & B & A \end{pmatrix}$ to the permutation $\begin{pmatrix} A & B & C & D \\ C & A & B & D \end{pmatrix}$, then we obtain $\begin{pmatrix} C & A & B & D \\ D & B & A & C \end{pmatrix}$, that is to say $\begin{pmatrix} A & B & C & D \\ B & A & D & C \end{pmatrix}$, which corresponds to the rotation through π about the axis PQ.

In the same way we convince ourselves that

$$-a + b + a$$

is the rotation about the axis joining the vertex A to the centroid of the face BCD and carrying C into B, D into C, and B into D.

[To this rotation there corresponds the permutation $\begin{pmatrix} A & B & C & D \\ A & C & D & B \end{pmatrix}$.]

3. Conjugate elements

Let G be any group.

Theorem I′.—If the element b is the transform of the element a by the element c, then a is the transform of b by —c.

Indeed it follows from

$$b = -c + a + c$$

by adding c to the left and $-c$ to the right of both sides, that

$$c + b + (-c) = a$$

and therefore

$$a = -(-c) + b + (-c)$$

Definition.—Two group elements are called *conjugate* if one of them is a transform of the other.

Theorem I″.—If a is conjugate to b, and b is conjugate to c, then also a is conjugate to c.

Since a is conjugate to b, there exists an element d such that

$$b = -d + a + d \tag{5}$$

since b is conjugate to c, there exists an element e such that

$$b = -e + c + e \tag{5'}$$

and hence

$$-d + a + d = -e + c + e$$

If we add d to the left of both sides of this last equation and $-d$ to the right of both sides, then we obtain

$$a = (d - e) + c + (e - d) = -(e - d) + c + (e - d)$$

i.e. a is the transform of c by $e - d$, which proves that a is conjugate to c.

Theorem I‴.—Every element is conjugate to itself.

Indeed

$$a = -0 + a + 0$$

Theorems I′, I″, I‴ assert that the conjugacy of two group elements is symmetrical, transitive, and reflexive.* From this it follows, according to Theorem III of the appendix, that

Theorem I.—Every group may be partitioned into classes of mutually conjugate elements.

The class of an arbitrary element a of the group G consists of all the elements of G which are conjugate to a, and therefore of the transforms of a by all possible elements of the group G.

We establish that the class of the null element of any group G consists of this element alone (since for arbitrary a we have $-a + 0 + a = 0$).

4. Transformation of a subgroup

The class of conjugate elements to which the element a of the group G belongs consists of the transforms of the element a *by all possible elements* b of the group G. We now choose an arbitrary subgroup H of G and we wish to consider the transforms of all possible elements x of this subgroup by one fixed, arbitrarily chosen, element b of the group G. The resulting set of elements, that is to say *the set of all elements of the form*

$$-b + x + b$$

where b is the particular element of the group G which we have chosen and x runs through all the elements of the subgroup H, is called the transform of the subgroup H by b; we denote it by

$$-b + H + b$$

Assertion: $-b + H + b$ is a group.

Proof.—1. Let c_1 and c_2 be two elements belonging to

$$-b + H + b$$

We show that $c_1 + c_2$ belongs to

$$-b + H + b$$

Now
$$\left.\begin{aligned} c_1 &= -b + x_1 + b \\ c_2 &= -b + x_2 + b \end{aligned}\right\} \tag{6}$$

where x_1 and x_2 are elements of the group H.

From (6) it follows immediately that

$$c_1 + c_2 = -b + x_1 + x_2 + b \tag{7}$$

* See Appendix § 5, in particular section 3.

and therefore $c_1 + c_2$ is the transform of the element $x_1 + x_2$ by b; hence $c_1 + c_2$ belongs to $-b + H + b$.

2. We show that the null element 0 of the group G belongs to $-b + H + b$. Indeed since 0 is an element of H and

$$-b + 0 + b = 0$$

then also 0 belongs to $-b + H + b$.

3. Finally if a belongs to $-b + H + b$ then so also does $-a$; for if a belongs to $-b + H + b$ then $a = -b + x + b$, where x is some element of H. But then

$$-a = -(-b + x + b) = -b + (-x) + b$$

and therefore $-a$ is the transform of the element $-x$ of the group H by b, and hence $-a$ is an element of the set $-b + H + b$.

Therefore $-b + H + b$ is a group.

To every element x of the group H there corresponds a uniquely determined element of the group $-b + H + b$, namely the element $-b + x + b$.

In this way to two distinct elements x_1 and x_2 there correspond two distinct elements $-b + x_1 + b$ and $-b + x_2 + b$; for if $x_1 \neq x_2$ then on account of the uniqueness of subtraction the elements $x_1 + b$ and $x_2 + b$ are also distinct,* and hence also the elements $-b + x_1 + b$ and $-b + x_2 + b$.† Therefore if we let the element x of the group H correspond to the element $-b + x + b$ of the group $-b + H + b$, then we obtain a one-to-one correspondence between H and $-b + H + b$. On account of the equations (6) and (7) there corresponds to the sum of two elements x_1 and x_2 the sum of the elements $-b + x_1 + b$ and $-b + x_2 + b$. Therefore this correspondence is an isomorphism between the groups H and $-b + H + b$.

We have therefore proved the following theorem:

Theorem II.—The transform of the subgroup H of the group G by an element b of G is itself a subgroup of G which is isomorphic to H.

Remark.—The following results are immediate consequences of the definition of the transform:

1. If G is a commutative group and H a subgroup, then the trans-

* From the relation $x_1 + b = x_2 + b = c$ it follows that $x_1 = c - b$ and $x_2 = c - b$.

† The relation $-b + x_1 + b = -b + x_2 + b = c$ gives rise to $x_1 + b = b + c$ and $x_2 + b = b + c$.

form of H by an arbitrary element b of G is H itself, since in this case the transform of any element x by b is x itself, i.e. $-b + x + b = x$.

2. If G is any group, H a subgroup of G, and b an element of H then

$$-b + H + b = H$$

Indeed, for any element x of H, the element $-b + x + b$ belongs to H since b belongs to H. Therefore

$$-b + H + b \subseteq H$$

Since $-b$ belongs to H we also have

$$-(-b) + H + (-b) \subseteq H$$

Whence

$$H \subseteq -b + H + b$$

Therefore

$$-b + H + b = H$$

If the subgroup H_2 is the transform of the subgroup H_1 by the element b then H_1 is the transform of H_2 by the element $-b$.

The proof follows from Theorem I′ of section 3.

Definition.—Two subgroups of a group G, one of which is a transform of the other, are called conjugate subgroups.

Since $-0 + H + 0 = H$ it follows that every group is conjugate to itself.

From Theorem I″ it follows that two subgroups which are conjugate to a third are also conjugate to each other, so that the set of all subgroups of a group G is partitioned into classes of mutually conjugate subgroups.

We know already (Theorem II of this section) that *all mutually conjugate subgroups are isomorphic to one another.*

5. Examples

As we have already seen, the rotation group of the regular tetrahedron has the following subgroups:

1. Two improper subgroups: Firstly the subgroup consisting of the null element alone, and secondly the subgroup consisting of all twelve rotations of the tetrahedron. Each of these subgroups is evidently conjugate only to itself.

2. Three subgroups of order 2: H_{01}, H_{02}, H_{03}, each of which consists of rotations through the angles 0 and π about a line joining the mid-

points of a pair of opposite sides of the tetrahedron. *All these groups form a class of conjugate subgroups.*

3. The group H of order 4 (Klein's four-group), which is the (set-theoretical) union of the three groups H_{01}, H_{02}, H_{03}, and which therefore consists of the identical rotation and the rotations through the angle π about the lines joining the mid-points of the three pairs of opposite sides. From the definition of the group H as the union of the groups H_{01}, H_{02}, H_{03}, and from the fact that the groups H_{01}, H_{02}, H_{03} form a class of conjugate subgroups, it follows that *the group H is conjugate only to itself.*

4. Four subgroups of order 3: H_0, H_1, H_2, H_3. Each of them consists of rotations through the angles 0, $2\pi/3$, $4\pi/3$ about a line joining a vertex to the centroid of the opposite face. *All these groups evidently form a class of conjugate subgroups.*

Therefore all ten subgroups of the rotation group of a regular tetrahedron may be divided up in the following way into classes of conjugate subgroups:

(*a*) three classes each consisting of a single element, namely the two classes each of which consists of one of the improper subgroups and the class consisting of the single subgroup of order 4,

(*b*) the class consisting of the three subgroups of order 2,

(*c*) the class consisting of the four subgroups of order 3.

§ 2. Invariant subgroups (normal divisors)

1. Definition

If a subgroup H of a given group G possesses no conjugate subgroups different from itself (if therefore the class of all subgroups which are conjugate in the group G to the subgroup H consists only of the group H) then we call the subgroup H an *invariant* subgroup* (or *normal divisor*) of the group G.

Evidently the definition of an invariant subgroup can also be formulated as follows:

We call a subgroup of a group G invariant if the transform of an arbitrary element of the group H by any element of the group G is again an element of the group H.

The idea of an invariant subgroup is one of the most important ideas in the whole of algebra. Even if it is impossible in this short

* The word "invariant" signifies that H is left unchanged by every transformation.

exposition to make clear to the reader the full significance of this concept, which is particularly apparent in algebra in connection with the so-called Galois theory, we hope that from the investigations of this chapter and the next the reader will see how very significant invariant subgroups are in the logical structure of group theory.

2. Examples

The two improper subgroups of any group are trivial examples of invariant subgroups. Moreover evidently every subgroup of a commutative group is an invariant subgroup.

We mention some less trivial examples.

1. The group of displacements of a straight line along itself is an invariant subgroup of the group of all congruences of the line (Chapter V, § 2).

2. The cyclic group A of order n which consists of all congruences of the first kind of an n-pointed double pyramid is an invariant subgroup of the group of all rotations of the double pyramid.*

3. The alternating permutation group A_n on n numbers is an invariant subgroup of the group S_n of all permutations on n numbers. For if b is an arbitrary element of the group A_n, and therefore an arbitrary even permutation, and if a is an arbitrary element of the group S_n, and therefore an arbitrary even or odd permutation, then the sign of the permutation $-a + b + a$ is equal to the product of three numbers, each of which is equal to $+1$ or -1:

$$(\text{sgn } -a) \,.\, (\text{sgn } b) \,.\, (\text{sgn } a)$$

Since $(\text{sgn } -a) = (\text{sgn } a)$, then $(\text{sgn } -a).\,(\text{sgn } a)$ is always equal to $+1$, that is to say equal to $+1$ for arbitrary a. It follows that

$$(\text{sgn } (-a + b + a)\,) = (\text{sgn } b) = +1$$

which means that $-a + b + a$ is an even permutation and therefore an element of the group A_n.

Hence the transform of an arbitrary element b of the group A_n is

* Indeed if a is a congruence of the first kind and b a congruence of the second kind then (as we have shown in Chapter V, § 3)

$$a + b = b - a$$

and hence

$$-b + a + b = -a$$

Since this is true for every element a of the group A, then

$$-b + A + b = A$$

again an element of the group A_n (in general different from b), i.e. A_n is an invariant subgroup of the group S_n.

We consider examples of both invariant and non-invariant subgroups.

We have already seen that in the group of all rotations of a tetrahedron there is one proper invariant subgroup which is of order 4. Since the group of all rotations of the tetrahedron is isomorphic to the alternating group A_4 of permutations on four elements (that is to say to the group of all even permutations on four elements), this result may also be formulated as follows: *The alternating permutation group on four elements possesses one invariant subgroup of order* 4.

This result is very important. *It turns out that for $n > 4$ the alternating permutation group A_n on n numbers contains no invariant subgroup* apart from its two improper subgroups. This fact, the proof of which the reader can find for example in *The Theory of Groups* by A. G. Kurosh, has great significance in algebra; it is closely connected with the result that in general an equation of degree $n > 4$ cannot be solved by radicals.

The rotation group of the cube is as we know isomorphic to the group S_4. Therefore it has an invariant subgroup isomorphic to A_4. This subgroup is already familiar to us from Chapter V, § 5. It consists of the rotations which carry over into themselves the two tetrahedra inscribed in the cube.

We have also already mentioned the three subgroups of order eight contained in the rotation group of the cube. These three groups form a class of mutually conjugate groups, and therefore none of them is invariant. However the intersection of these three groups, consisting as we know of the null element and of three rotations each through an angle π about a line joining the mid-points of a pair of opposite faces, is an invariant subgroup.*

The rotation group of the cube possesses no proper invariant subgroups other than the above-mentioned groups of orders 12 and 4.

We mention further the following classes of conjugate groups:

1. The class consisting of three cyclic groups of order 4; each of these groups consists of rotations about one of the axes joining the mid-points of two opposite faces of the cube.

2. The class consisting of four cyclic groups of order 3; each of these groups consists of rotations about one of the diagonals.

3. The class consisting of six cyclic groups of order 2; each of

* The reader is advised to prove the following general theorem: The intersection of all groups belonging to a certain class of mutually conjugate subgroups is an invariant subgroup.

these groups consists of rotations about one of the axes joining the mid-points of two opposite edges.

Finally we consider more closely the already familiar group of movements of a plane in itself (Chapter V, § 2).

We make the following preliminary remark. Every movement of the plane in itself associates with every point x of the plane a certain uniquely determined point $f(x)$ of the plane, namely that point into which the point x is carried by the given movement.

We can therefore regard every movement as a certain mapping of the plane on itself. This mapping is a congruent mapping in the sense that the distance between points is left unchanged. If the two points x and y are carried over into $f(x)$ and $f(y)$, then the distance between $f(x)$ and $f(y)$ is equal to the distance between x and y. From this it follows in particular that two different points of the plane can never be mapped into one and the same point. If the two points x and y are different, then the distance between them is not equal to zero. But then the distance between the points $f(x)$ and $f(y)$ must also be different from zero; therefore the points $f(x)$ and $f(y)$ cannot coincide. *Therefore every movement is a one-to-one mapping of the plane on itself.*

A movement, interpreted as a one-to-one mapping of the plane on itself, will be denoted by the symbol $f(x)$, where x is of course a general point of the plane.

Suppose that two movements $f(x)$ and $\varphi(x)$ are given. We wish to construct the transform of $f(x)$ by $\varphi(x)$. By definition this is the movement

$$F(x) = \varphi\{f[\varphi^{-1}(x)]\} \tag{1}$$

Since $\varphi(x)$ is a one-to-one mapping of the plane the movement $F(x)$ is completely determined when we know into what point $\varphi(x)$ is carried, for arbitrary x, by this movement. In other words, the mapping $F(x)$ is defined for arbitrary x when we know into what point $\varphi(x)$ is carried (again for arbitrary x). If we now replace x by $\varphi(x)$ in the formula (1) and observe that $\varphi^{-1}[\varphi(x)] = x$ then we obtain

$$F[\varphi(x)] = \varphi[f(x)] \tag{2}$$

The movement $F(x)$ is completely determined by this formula.

If we put

$$f(x) = y$$

then formula (2) may be expressed as follows:

For arbitrary x the movement F carries the point $\varphi(x)$ over into the point $\varphi(y)$.

We prove now the following assertion:

If f is a rotation through the angle α about the point a, then F is a rotation through the same angle α about the point $\varphi(a)$ (fig. 17).

Since f is a rotation about a, then

$$f(a) = a$$

from which it follows by formula (2) that

$$F[\varphi(a)] = \varphi(a)$$

and therefore F is a rotation about $\varphi(a)$. The movement f rotates an arbitrary half-line h issuing from a through the angle α and carries it over into the half-line $f(h)$. The movement φ which is a congruent

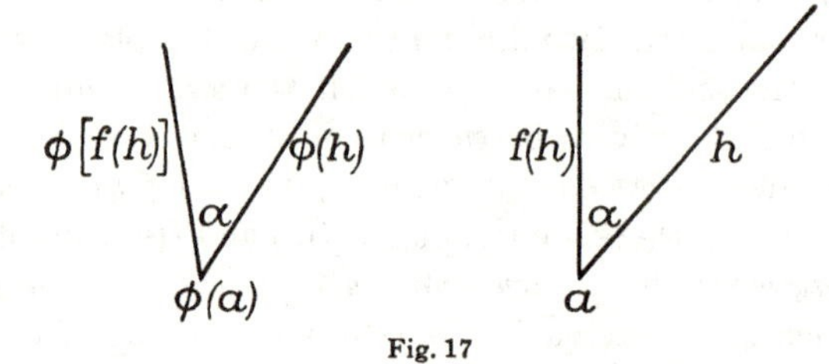

Fig. 17

mapping carries over the figure consisting of the two half-lines h and $f(h)$ issuing from a and inclined at an angle α, into a congruent figure consisting of the two half-lines $\varphi(h)$ and $\varphi[f(h)] = F[\varphi(h)]$ issuing from $\varphi(a)$. Therefore the half-line $F[\varphi(h)]$ is obtained from the half-line $\varphi(h)$ by a rotation through the angle α; i.e. since the movement F rotates the half-line $\varphi(h)$ through the angle α, F is a rotation through α.

From what we have just proved it follows that:

The transform, by an arbitrary movement φ, of the group of rotations of the plane about a point a is the group of rotations of the plane about the point $\varphi(a)$.

Let f be a translation of the plane along the line g and let φ be an arbitrary movement of the plane.

Then first of all we have the identity

$$f(g) = g$$

i.e. the line g goes over into itself by the movement f.

The movement φ carries the line g over into the line $\varphi(g)$. Applying formula (2) to each point of the line g it follows that

$$F[\varphi(g)] = \varphi(g)$$

The movement F therefore carries over the line $\varphi(g)$ into itself and hence is a translation along this line. Since φ is a congruent mapping the distance between x and $f(x)$ is equal to the distance between $\varphi(x)$ and $\varphi[f(x)]$, and therefore between $\varphi(x)$ and $F[\varphi(x)]$.

This means that *the translation F displaces the points of the plane through the same distance as does the translation f.*

From what we have proved it follows that:

The group of parallel-displacements of the plane along a given line g is transformed by any given movement φ into the group of parallel-displacements of the plane along the line $\varphi(g)$.

Since every movement φ transforms every parallel-displacement of the plane into a parallel-displacement we obtain the following important result:

The group of all parallel-displacements of the plane (along all possible lines) is an invariant subgroup of the group of all movements of the plane in itself.

EXERCISES ON CHAPTER VI

1. Find the transform b of $a = \begin{pmatrix} 1 & 2 & 3 & 4 \\ 1 & 3 & 4 & 2 \end{pmatrix}$ by $c = \begin{pmatrix} 1 & 2 & 3 & 4 \\ 2 & 3 & 1 & 4 \end{pmatrix}$, and verify that the transform of b by $-c$ is a.

2. Find the transform of the element b in Ex. 1 by the element a. Show that this element is conjugate to a in S_4.

3. Prove that the rotation group of the tetrahedron is partitioned into the following classes of conjugate elements:

(1) the class consisting of the null element alone;

(2) the class which consists of the rotations through the angle $2\pi/3$ about each of the four axes joining a vertex of the tetrahedron to the centroid of the opposite face;

(3) the class which consists of the four rotations through the angle $4\pi/3$ about the same axes (here and in (2) the rotations are measured either in the clockwise or counter-clockwise sense when viewed from the fixed vertex);

(4) the class which consists of the rotations through the angle π about the three axes joining the mid-points of the pairs of opposite sides of the tetrahedron.

4. Determine the classes of conjugate elements in the symmetric group S_3 and the classes of conjugate subgroups.

5. Determine the classes of conjugate subgroups in the dihedral group of order 8, and hence its invariant subgroups.

6. The symmetric group S_n of permutations on the n numbers $1, 2, \ldots, n$ is evidently a subgroup of the symmetric group S_m of permutations on the m numbers $1, 2, \ldots, m$ for $m > n$. Prove that this subgroup is never invariant.

7. Prove that the set of elements which commute with every element of a group G form an invariant subgroup of G. (It is called the *centre* of G.)

8. Prove that the centre of S_3 consists only of the identical permutation.

9. Find the centre of the dihedral group of order 8.

Chapter VII

HOMOMORPHIC MAPPINGS

§ 1. Definition of a homomorphic mapping and its kernel

Definition and simple properties

We suppose that there is associated with each element a of a group A an element

$$b = f(a)$$

of a group B. The totality of elements $b = f(a)$ of the group B so obtained we shall denote by $f(A)$. We say that what we are considering is a mapping of the group A *into* the group B; that is to say, in set notation, $f(A) \subseteq B$.

We introduce the following fundamental definition:

A mapping f of a group A into a group B is called homomorphic when the condition

$$f(a_1 + a_2) = f(a_1) + f(a_2) \tag{1}$$

is fulfilled for any pair of elements a_1 and a_2 of the group A, where the sign $+$ must naturally be understood on the left-hand side of equation (1) as the sign of addition in the group A, whereas on the right-hand side as the sign of addition in the group B.

Theorem.—If f is a homomorphic mapping of a group A into a group B, then the set $f(A) \subseteq B$ is a subgroup of the group B.

Proof.—It is sufficient to prove

1. If b_1 and b_2 are elements of the set $f(A)$, then likewise $b_1 + b_2$ is an element of the set $f(A)$.

2. The null element of the group B is an element of the set $f(A)$.

3. If b is an element of the set $f(A)$, likewise $-b$ is an element of the set $f(A)$.

We prove these steps 1, 2, 3 in turn.

1. Let b_1 and b_2 be two elements of the set $f(A)$. This signifies the existence of elements a_1 and a_2 of the group A satisfying

$$f(a_1) = b_1 \text{ and } f(a_2) = b_2$$

But since the mapping f is homomorphic, we have

$$f(a_1 + a_2) = b_1 + b_2$$

Accordingly, by means of the mapping f, the element $a_1 + a_2$ of the group A corresponds to the element $b_1 + b_2$ of the set $f(A)$. The first step is therefore proved.

2. Let 0 be the null element and a any other element of the group A. In the group A we have

$$a + 0 = a$$

from which it follows that for the group B

$$f(a + 0) = f(a)$$

But since the mapping is homomorphic we have

$$f(a) + f(0) = f(a)$$

i.e. $f(0)$ is the null element of the group B. This finishes the second step.

3. Let b be any element of the set $f(A) \subseteq B$. There exists an element a of the group A such that

$$f(a) = b$$

We denote by b' the element $f(-a)$ of the set $f(A)$, and prove that

$$b' = -b$$

Now we have

$$a + (-a) = 0$$

thus it follows that

$$f(a) + f(-a) = 0'$$

($0'$ denoting the null element of the group B), and therefore

$$b + b' = 0'$$

i.e.

$$b' = -b$$

which is what was to be proved.

Consequently each homomorphic mapping of a group A into a group B is a homomorphic mapping of the group A onto a certain subgroup of the group B.

Remark I.—Two important assertions, which are true for every homomorphic mapping of a group A into a group B, are contained in the discussion just carried through, namely

$$f(0) = 0' \tag{2}$$

and

$$f(-a) = -f(a) \tag{3}$$

Remark II.—In view of the fundamental remark in Chapter III, § 2, we can state that:

A one-to-one homomorphic mapping of a group A onto a group B is an isomorphic mapping.

Definition.—Let f be a homomorphic mapping of a group A into a group B. The set of all elements x of the group A that are mapped by f into the null element of the group B is called the *kernel* of the homomorphic mapping f and is denoted by $f^{-1}(0')$.

Theorem.—*The kernel of a homomorphic mapping f of a group A into a group B is an invariant subgroup of the group A.*

Proof.—From the definition of a homomorphic mapping it follows that the equations

$$f(a_1) = 0', \quad f(a_2) = 0'$$

imply

$$f(a_1 + a_2) = 0'$$

Thus if a_1 and a_2 are elements of $f^{-1}(0')$, then $a_1 + a_2$ is also an element of $f^{-1}(0')$.

Further we have seen from the proof of the previous theorem that $f(0)$ is the null element of the group B; thus 0 is an element of $f^{-1}(0')$.

Finally if $f(a) = 0'$, then $f(-a) = -f(a) = 0'$, and we conclude that if a is an element of $f^{-1}(0')$, then so is $-a$. From this it follows readily that $f^{-1}(0')$ is a subgroup of the group A.

In order to prove that $f^{-1}(0')$ is an invariant subgroup of the group A, we must convince ourselves that the transform $-a + x + a$ of any element x of the group $f^{-1}(0')$ with respect to any element a of the group A is again an element of the group $f^{-1}(0')$. In other words, we must convince ourselves that we have

$$f(-a + x + a) = 0'$$

whenever $f(x) = 0'$.

But this is immediately obvious, for if $f(x) = 0'$ then

$$f(-a + x + a) = -f(a) + f(x) + f(a)$$
$$= -f(a) + 0' + f(a) = -f(a) + f(a) = 0'$$

This completes the proof of the theorem.

We shall see later that conversely each invariant subgroup of a group A is a kernel of a certain homomorphic mapping of the group A.

§ 2. Examples of homomorphic mappings

I. We consider the group G of all whole numbers

$$\ldots, -n, -(n-1), \ldots, -2, -1, 0, 1, 2, \ldots, (n-1), n, \ldots$$

and a group G_2 of order two whose elements are b_0 and b_1 and whose addition table is accordingly

$$b_0 + b_0 = b_0, \quad b_0 + b_1 = b_1 + b_0 = b_1, \quad b_1 + b_1 = b_0$$

Obviously, b_0 is the null element of the group G_2.

We now construct the following mapping f of the group G onto the group G_2:

With each even number we associate the element b_0 of the group G_2 and with each odd number the element b_1 of the group G_2. This mapping is homomorphic. For let a and a' be two whole numbers. If both a and a' are even numbers then $a + a'$ is likewise even, and we have

$$f(a + a') = f(a) = f(a') = b_0 = f(a) + f(a')$$

If one of the two numbers a and a', say a, is even while the other is odd, then $a + a'$ is odd, so that we have

$$f(a) = b_0 \qquad f(a') = b_1$$
$$f(a + a') = b_1 = b_0 + b_1 = f(a) + f(a')$$

If finally a and a' are both odd numbers, then $a + a'$ is an even number and we have

$$f(a) = f(a') = b_1, \quad f(a + a') = b_0 = b_1 + b_1 = f(a) + f(a')$$

The kernel of our homomorphism is obviously the group of all even numbers.

We generalize this example. Let any natural number $m \geqq 2$ be given. We examine the cyclic group G_m of order m with the elements $b_0, b_1, b_2, \ldots, b_{m-1}$ and the addition table

	b_0	b_1	b_2	...	b_{m-2}	b_{m-1}
b_0	b_0	b_1	b_2	...	b_{m-2}	b_{m-1}
b_1	b_1	b_2	b_3	...	b_{m-1}	b_0
b_2	b_2	b_3	b_4	...	b_0	b_1
...	...	...	...	...	...	...
b_{m-2}	b_{m-2}	b_{m-1}	b_0	...	b_{m-4}	b_{m-3}
b_{m-1}	b_{m-1}	b_0	b_1	...	b_{m-3}	b_{m-2}

(the null element is denoted by b_0).

We now construct a homomorphic mapping f of the group G of all whole numbers onto the group G_m.

For this purpose we call to mind beforehand the following theorem of arithmetic: *Each whole number a upon division by a natural number m leaves as remainder one of the numbers* $0, 1, \ldots, m - 1$. *This means that the remainder corresponding to the number a is defined as the uniquely determined non-negative number r, which satisfies the conditions*

$$a = mq + r, \qquad 0 \leqq r \leqq m - 1 \tag{1}$$

where q is a whole number (q is called the quotient of the division of a by m). This theorem is generally well known for positive a. For $a = 0$, we obviously have

$$0 = m \,.\, 0 + 0$$

so on division of 0 by any natural number we obtain zero for the quotient and also for the remainder.

The case of a negative a however requires, perhaps, some explanation. If a is negative, then $-a$ is positive.

We divide the natural number $-a$ by the natural number m, denoting the quotient by q' and the remainder by r'. We may assume

r' to be positive (should $r' = 0$, then $-a$ and also consequently a would be without remainder when divided by m). Thus we have

$$-a = mq' + r' \quad (0 < r' \leqq m - 1)$$

and consequently

$$a = -mq' - r' = -m - mq' + m - r' = m(-1 - q') + (m - r')$$

From $0 < r' \leqq m - 1$ it obviously follows that

$$0 < m - r' \leqq m - 1$$

If we set $q = -1 - q'$, $r = m - r'$, then we have for the whole numbers a, q, r the relation

$$a = mq + r \quad (0 \leqq r \leqq m - 1) \tag{2}$$

We easily convince ourselves that the representation of the whole number a in equation (2) for a given natural number m is *unique*, q, r being integral and $0 \leqq r \leqq m - 1$; that is, the whole numbers q and r are completely defined by the conditions (2). For suppose we also have

$$a = mq_1 + r_1 \quad (0 \leqq r_1 \leqq m - 1) \tag{2'}$$

Then if we subtract equation (2′) term by term from equation (2) we obtain

$$0 = m(q - q_1) + r - r_1$$

i.e.

$$r - r_1 = m(q_1 - q)$$

From this it follows that the whole number $r - r_1$ is without remainder when divided by m. But $r - r_1$ is the difference of two non-negative numbers that are not greater than $m - 1$, hence the absolute value of this difference is also not greater than $m - 1$. Therefore the number $r - r_1$ can only be without a remainder when divided by m when it is equal to zero. Thus we have

$$r - r_1 = 0 \quad r = r_1$$

and

$$a = mq_1 + r \tag{3}$$

From equations (3) and (2) we obtain

$$q_1 = \frac{a - r}{m} \qquad q = \frac{a - r}{m}$$

$$q_1 = q$$

which is what was to be proved.

In consequence of the inequality

$$0 \leqq r \leqq m - 1$$

there corresponds to the whole number r the element b_r of the group G_m. For any fixed chosen natural number $m \geqq 2$, there thus corresponds to each whole number a, a uniquely determined element of the cyclic group G_m of order m, namely the element b_r where r is the remainder when a is divided by m. *We shall call this element b_r the remainder of the number a modulo m.*

By means of the relation just stated, a mapping f of the group G onto the group G_m is generated. We prove that this mapping f is homomorphic.

Let a and a' be two whole numbers and let

$$\left.\begin{aligned} a &= mq + r \qquad & 0 \leqq r \leqq m - 1 \\ a' &= mq' + r' \qquad & 0 \leqq r' \leqq m - 1 \end{aligned}\right\} \qquad (4)$$

Then we have

$$a + a' = m(q + q') + r + r'$$

But now the number $r + r'$, which naturally satisfies the inequality $0 \leqq r + r'$, need not satisfy the inequality $r + r' \leqq m - 1$. But certainly we have

$$r + r' = mq'' + \rho$$

where q'' is the quotient in the division of $r + r'$ by m (we easily see that it is either equal to 0 or 1) and ρ the remainder in this division; therefore we have

$$a + a' = m(q + q' + q'') + \rho \quad (0 \leqq \rho \leqq m - 1)$$

Thus to the element $a + a'$ there corresponds in our mapping f the element b_ρ of the group G_m.

If we examine the addition table of the cyclic group of order m we see that

$$b_r + b_{r'} = b_\rho$$

(where ρ as formerly is the remainder in the division of $r + r'$ by m). Thus we have

$$f(a + a') = b_\rho = b_r + b_{r'} = f(a) + f(a')$$

which proves that the mapping f is homomorphic.

The construction that we have just accomplished of a homomorphic mapping f of the group of all whole numbers onto the cyclic group of order m is of fundamental importance in the elementary theory of numbers. We shall denote this homomorphic mapping by f_m.

The kernel of the homomorphism f_m is the group of all whole numbers that have no remainder when divided by m.

II. Let A be the group of all movements of a plane in itself. We choose in the plane a fixed point O and a fixed vector h issuing from O. Each movement f of the plane in itself carries over the vector h into a vector $f(h)$. The vector $f(h)$ forms with the vector h a certain angle* which we denote by ω_f. This angle is zero if and only if the vectors $f(h)$ and h are parallel and in the same sense, in which case the movement f is therefore a parallel displacement.

Now we associate with the movement f a rotation of the plane through the angle ω_f. In this way we obtain a mapping of the group of all movements of the plane onto the group of all rotations of the plane about the point O, and onto the group κ isomorphic to it (see Chapter V, § 2). This mapping is homomorphic, as the reader can easily convince himself. The kernel of this mapping is the group of parallel displacements of the plane.

III. In Chapter V, § 2, in the second example it was shown that to each real number there corresponds a member of the group κ. By means of this correspondence a homomorphic mapping of the group of all real numbers onto the group κ is produced, and the kernel of this mapping is the infinite cyclic group, consisting of all real numbers which are integral multiples of 2π.

EXERCISES ON CHAPTER VII

1. Show that the permutations of the forms $\begin{pmatrix} 1 & 2 & 3 & 4 & \dots & k \\ 1 & 2 & a_3 & a_4 & \dots & a_k \end{pmatrix}$ and $\begin{pmatrix} 1 & 2 & 3 & 4 & \dots & k \\ 2 & 1 & b_3 & b_4 & \dots & b_k \end{pmatrix}$, where $a_3, a_4, \dots, a_k$ and $b_3, b_4, \dots, b_k$ are rearrangements of $3, 4, \dots, k$, form a subgroup of order $2! \times (k-2)!$ of the symmetric group S_k.

2. (*a*) Show that the subgroup of Ex. 1 can be mapped homomorphically onto S_2.

(*b*) Consider the analogous problem of a subgroup of S_k of order $3! \times (k-3)!$ mapped homomorphically onto S_3.

3. What are the kernels of the homomorphisms of Ex. 2? Verify by the methods of Chapter VI that they are invariant subgroups of the appropriate groups.

4. Show that if a cyclic group is mapped homomorphically onto a group G, then G must also be cyclic.

* We obtain this angle between the vector h and the vector $f(h)$ if we draw through the point O the vector that is parallel to and in the same sense as the vector $f(h)$.

Chapter VIII

PARTITIONING OF A GROUP RELATIVE TO A GIVEN SUBGROUP

DIFFERENCE MODULES

§ 1. Left and right cosets

1. Left cosets

Let a group G be given and in it a subgroup U. We now set ourselves the task of proving the following: The given subgroup U defines (and indeed in general in two different ways) a division of the group G into a certain system of mutually disjoint subsets, one of which is the subgroup U itself, while the remaining ones can be constructed uniquely from U by means of an extremely simple procedure.

In order to obtain this partition we proceed as follows: We call two elements a and b of the group G *equivalent* with respect to the subgroup U, when the left difference of the elements b and a, that is to say the element $-a+b$, is an element of the subgroup U.

This property of equivalence (we call it *left-sided equivalence*) is *symmetric*. In fact if

$$-a+b=u$$

where u is an element of the group U, then

$$-b+a=-(-a+b)=-u$$

and $-u$ is likewise an element of the subgroup U.

This equivalence is *transitive*. If namely

$$-a+b=u_1$$
$$-b+c=u_2$$

where u_1 and u_2 are elements of the subgroup U, then

$$-a+c=(-a+b)+(-b+c)=u_1+u_2$$

and u_1+u_2 is likewise an element of the subgroup U.

Finally, this equivalence is *reflexive*, since

$$-a + a = 0$$

and 0 is an element of the subgroup U.

Thus on the basis of Theorem III of § 5 of the appendix the group G is partitioned into sets of elements which are equivalent to each other relative to the subgroup U. *These sets are called left cosets of the group G with respect to the subgroup U.* We point out that the left coset $'K_a$ of an element a of a group G (i.e. the left coset containing a) consists of all elements x satisfying the relationship $-a + x = u$, where u is an element of the subgroup U, i.e. *of all elements of the form $x = a + u$, where u is an element of the subgroup U.*

We remark further that if a is an element of U (in particular if $a = 0$) then $'K_a = U$, because in this case $a + u$ is an element of U for any u from the group U; and each element u of the group U can be exhibited in the form $a + u_1$, where again $u_1 = -a + u$ denotes an element of the group U. Since each element of the set $'K_a$ can be represented in the form $a + u$, and since for different elements u_1 and u_2 of the group U the elements $a + u_1$ and $a + u_2$ of the set $'K_a$ are different, we therefore obtain *a one-to-one correspondence between U and any $'K_a$*, if we make correspond to each element u of the group U *the element $a + u$ of the set $'K_a$.*

Finally we remark that *among all the sets $'K_a$ only one set is a subgroup of G, namely U.*

Indeed if $'K_a$ is a subgroup, then the null element of the group G must belong to $'K_a$. It follows that it is a common element of the sets $'K_a$ and U, and hence $'K_a$ coincides with U.

2. The case of a finite group G

Because of the one-to-one correspondence existing between $'K_a$ and the subgroup U, in a finite group G all the $'K_a$ have an equal number of elements, say m, where m is the order of the group U. If the number of different sets is equal to j, and n is the order of the group G, then we obviously have $n = mj$.

From this there follows in particular a result which was mentioned earlier (Chapter II, § 2) namely

Lagrange's Theorem.—The order of each subgroup of a finite group G is a divisor of the order of the group G.

The number j, that is to say the number of left cosets* of the

* This number can also be finite in the case of an infinite group G. Thus for example, when the group G is the group of all whole numbers, and U is that subgroup of G which consists of all numbers that are exactly divisible by the whole number $\mu \geqq 2$.

group G with respect to the subgroup U, is called *the index of a subgroup U in the group G.*

3. Right cosets

We now call two elements a and b equivalent (*right-sided equivalence*) relative to the subgroup U, when their right difference

$$b - a = b + (-a)$$

is an element of the subgroup U. We easily verify that this equivalence is symmetric, transitive, and reflexive.

Indeed it follows from

$$b - a = u$$

where u is an element of the group U, that

$$a - b = -(b - a) = -u$$

and it follows from

$$b - a = u_1 \quad c - b = u_2$$

where u_1 and u_2 belong to U, that

$$c - a = (c - b) + (b - a) = u_1 + u_2$$

Finally

$$a - a = 0$$

and 0 belongs to U.

The right-equivalence defines a partition of the group G into *right* cosets, where the *right coset K'_a of the given element a consists of all elements x, for which $x - a = u$ is an element of the group U*, and thus of all elements of the form

$$x = u + a$$

where u belongs to U.

If a belongs to U, then the set K'_a coincides with U.

If we let each element u of the subgroup U correspond to the element $u + a$ of the coset K'_a then we obtain a one-to-one correspondence between U and the set K'_a. In the case of a finite subgroup U, all cosets K'_a of this subgroup are finite and consist of the same number of elements as U itself. If the group G is finite of order n, and the subgroup U has order m, we have as before

$$n = mj$$

where j is the number of different right cosets of the subgroup U, which is therefore equal to the number of different left cosets.

Thus the index of a subgroup U relative to a group G can be defined equally well as the number of left cosets or as the number of right cosets of the group G with respect to the subgroup U. It is equal to the order of G divided by the order of U.

4. The coincidence of the left and right cosets in the case of an invariant subgroup

The question now arises under what circumstances can we have

$$'K_a = K'_a$$

for each element a of the group G.

Obviously for this it is necessary and sufficient that each element of the form $a + u$ is equal to a certain element $u' + a$, and conversely that each element $u + a$ is equal to a certain element $a + u'$ (here u, u' always denote elements of the subgroup U). Both conditions are equivalent; for the first condition states that for each a of G and each u of U we can find a u' of U in such a way that we have

$$a + u = u' + a$$

whence

$$a + u + (-a) = u'$$

This implies

$$-(-a) + U + (-a) = U$$

Since any one element of the group G can be represented in the form $-a$ for a suitable choice of the element a the first condition simply means that the transform of the subgroup U with respect to any element of the group G coincides with U, or *U is an invariant subgroup of the group G.*

The second condition reads: For each a of G and each u of U we can find a u' from U in such a way that

$$u + a = a + u'$$

and therefore

$$-a + u + a = u'$$

This implies

$$-a + U + a = U$$

Consequently the second condition likewise requires that U shall be an invariant subgroup of the group G.

Thus we have proved this theorem:

Theorem.—Let U be a subgroup of the group G. For every element a of the group G the left coset of this element with respect to the subgroup U

coincides with the right coset of the same element if and only if U is an invariant subgroup of the group G.

Since for an invariant subgroup U we have for each element a of the group G

$$'K_a = K'_a$$

thus in place of $'K_a$ and K'_a we can simply write $K_a = 'K_a = K'_a$ and call this set simply *the coset of the element a with respect to the invariant subgroup U.*

In particular the right cosets coincide with the left when U is a subgroup of a commutative group G, because all subgroups of a commutative group are normal divisors (Chapter VI, § 2, section 2).

5. Examples

I. Let G be the group of all whole numbers and $U \subseteq G$ the group of all of the numbers that are without remainder when divided by m.

If a is any whole number, then K_a consists of all numbers of the form $a + mq$ with integral q. These are all the numbers which on division by m yield the same remainder as does the number a. Therefore the number of distinct cosets is equal to the number of different remainders that occur after division by m. But this number is equal to m, because the numbers $0, 1, 2, \ldots, m - 1$, and only these, occur as remainders on division by m. Thus we have the following cosets:

(0) The set of all numbers that on division by m yield the remainder 0. This coincides with the group U and consists of the numbers

$$\ldots, -qm, -(q-1)m, \ldots,$$
$$-3m, -2m, -m, 0, m, 2m, 3m, \ldots, qm, \ldots$$

(1) The set of all numbers that on division by m yield the remainder 1. These are

$$\ldots, -qm + 1, -(q-1)m + 1, \ldots, -3m + 1, -2m + 1,$$
$$-m + 1, 1, m + 1, 2m + 1, 3m + 1, \ldots, qm + 1, \ldots$$

(2) The set of all numbers that on division by m yield the remainder 2. These are the numbers

$$\ldots, -qm + 2, -(q-1)m + 2, \ldots, -3m + 2, -2m + 2,$$
$$-m + 2, 2, m + 2, \ldots, qm + 2, \ldots$$

. . . .

$(m-1)$ The set of all numbers that on division by m yield the remainder $(m-1)$. This set consists of the numbers

$$-qm+(m-1),\ -(q-1)m+(m-1), \dots,$$
$$-3m+(m-1),\ -2m+(m-1),\ -m+(m-1)$$
$$(m-1),\ m+(m-1),\ 2m+(m-1), \dots,\ qm+(m-1), \dots$$

or, what is the same thing, the numbers

$$\dots, -2m-1,\ -m-1,\ -1,\ m-1,\ 2m-1,\ 3m-1, \dots$$

II. Let G be the group S_3 of all permutations on three elements and U the subgroup of order 2 (and consequently of index 3), that consists of the following permutations:

$$P_0=\begin{pmatrix}1 & 2 & 3\\ 1 & 2 & 3\end{pmatrix} \quad \text{and} \quad P_2=\begin{pmatrix}1 & 2 & 3\\ 2 & 1 & 3\end{pmatrix}$$

The division of the group G into left and right cosets is evident from the following table:

Left cosets	Right cosets
$U=(P_0, P_2)$	$U=(P_0, P_2)$
(P_1, P_3)	(P_1, P_4)
(P_4, P_5)	(P_3, P_5)

III. The alternating permutation group A_n on n elements is itself an invariant subgroup of index 2 of the symmetric group S_n. The two cosets that belong to this subgroup are the group A_n itself and the set of all odd permutations.

IV. In the group of rotations of an n-pointed double pyramid the congruences of the first kind form an invariant subgroup of index 2. One of the two cosets of this subgroup is itself and the other consists of all congruences of the second kind.

V. The group U of all translations of a line along itself is an invariant subgroup of index 2 in the group G of all congruences of the line. The two cosets defined by this subgroup are the group U itself and the set of all congruences of the second kind.

VI. Let G be the group of all complex numbers with the usual addition as the group operation. Let U be the subgroup of all real numbers. The cosets into which the commutative group G is partitioned relative to the subgroup U are the sets K_β each of which consists of all complex numbers of the form

$$x + i\beta$$

where x and β are real numbers, β is given and x runs through the set of all real numbers. If we let the complex numbers correspond* to points of the plane in the usual way, then each set appears as a line parallel to the real axis (that is to say to the x-axis).

§ 2. The difference module corresponding to a given invariant subgroup

1. Definition

Let U be an invariant subgroup of a certain given group G. We consider the set of all cosets into which the group G is divided relative to the subgroup U. We denote this set by V and prove that we can define in it a law of addition so that V becomes a group onto which the group G can be mapped homomorphically.

Let v_1 and v_2 be two arbitrary elements of the set V. Then v_1 and v_2 define two cosets of the group G with respect to the invariant subgroup U. We choose a certain element in each of these cosets, say an element x_1 in the set v_1 and an element x_2 in the set v_2. We shall denote by v_3 the coset containing the element $x_1 + x_2$ of the group G.

We prove that the set v_3 does not depend on the particular elements x_1 and x_2 chosen from the sets v_1 and v_2. In other words, we prove that if x_1' is any element of the set v_1 which is in general different from x_1, and x_2' any element of the set v_2 which is in general different from x_2, then the element $x_1' + x_2'$ lies in the same coset v_3 as does the element $x_1 + x_2$.

In fact two elements a and b belong to the same coset of the invariant subgroup U if and only if their difference belongs to U.

We consider the difference

$$\begin{aligned}(x_1 + x_2) - (x_1' + x_2') &= x_1 + x_2 - x_2' - x_1' \\ &= x_1 + (x_2 - x_2') - x_1'\end{aligned}$$

* We are speaking here of the representation of the complex number $x + iy$ by the point in the plane with coordinates x and y.

Since x_2 and x_2' belong to one and the same coset v_2 we have

$$x_2 - x_2' = u_2$$

where u_2 is a certain element of U; this gives

$$(x_1 + x_2) - (x_1' + x_2') = x_1 + u_2 - x_1' \tag{1}$$

But U is an invariant subgroup, and hence

$$x_1 + u_2 = u' + x_1$$

where u' is an appropriate element of the group U.

If we substitute this in formula (1) then we obtain

$$(x_1 + x_2) - (x_1' + x_2') = u' + x_1 - x_1'$$

Now x_1 and x_1' belong to the same coset v_1 and hence $x_1 - x_1' = u_1$, where u_1 is a certain element of the group U. Consequently we have

$$(x_1 + x_2) - (x_1' + x_2') = u' + u_1$$

i.e. $(x_1 + x_2) - (x_1' + x_2')$ is an element $u = u' + u_1$ of the group U, which is what was to be proved.

Because the set v_3 so obtained is defined as soon as v_1 and v_2 are defined we may write

$$v_1 + v_2 = v_3 \tag{2}$$

This is to be understood as the *definition* of the sum $v_1 + v_2$ of two cosets v_1 and v_2.

Thus

The sum of two cosets v_1 and v_2 means that coset v_3 which is constructed according to the following rule:

In each set v_1 and v_2 we choose an arbitrary element, we add these two elements to each other, and we find the coset to which their sum belongs. This is the set v_3.

From this definition and from the fact that the addition of the elements of the group G satisfies the associative law, it follows immediately that the addition of cosets satisfies the associative law.

We prove that the set U, with respect to the law of addition just defined, plays the role of the null element, so that therefore we have for each coset v the equation

$$v + U = U + v = v \tag{3}$$

To this end we select an arbitrary element x from the set v and take the null element 0 from the set U. Then, from the definition of addition, it follows that the set $v + U$ is the coset containing the element $x + 0 = x$, that is to say it is the set v. Likewise the set $U + v$ is the coset containing $0 + x = x$, and is therefore the same set v. This proves formula (3).

Finally we prove that there exists a coset inverse to each co-set K, which we denote by $-K$, and which satisfies the condition

$$K + (-K) = (-K) + K = U$$

For this purpose we choose some element a in the set K and define the set $-K$ as that coset containing the element $-a$. From the definition of the addition of cosets it follows that each of the sums $K + (-K)$ and $(-K) + K$ represents the coset to which the element $a + (-a) = (-a) + a = 0$ belongs, and this is the set U.

Thus our definition of addition satisfies all the group axioms. *Consequently with respect to our definition of addition the aggregate of cosets of the group relative to one of its invariant subgroups U is a certain group V. The set U is moreover the null element of the group V.*

The group V is called the *difference module of the group G relative to its invariant subgroup U (it is denoted by $G - U$).**

2. The homomorphism theorem †

As before let a group G and one of its invariant subgroups U be given. With each element x of the group G we associate a certain element of the difference module V, namely the coset that contains the element x. From the mapping φ of the group G onto the group V thus constructed and from the definition of addition in the group V it follows immediately that this mapping is homomorphic.

Which elements of the group G will be mapped on the null element of the group V? Since the null element is U, the obvious answer to our question is that all elements of the invariant subgroup U, and only those, are mapped by φ on the null element of the group V.

From the investigations of this and the preceding section it follows that each invariant subgroup U of the group G is the kernel of a certain homomorphic mapping of the group G, namely the homomorphic mapping of the group G onto its difference module with respect to U.

Let us now consider an arbitrary homomorphic mapping f of a group A onto a group B. Let U be the kernel of this homomorphic

* When the group operation is represented multiplicatively, then we call V a *factor group* and denote it by G/U.

† See appendix, § 5, section 2.

mapping. We know that U is an invariant subgroup of the group A. We denote by V the difference module of the group A relative to U.

Let b be some element of the group B. Then there exists at least one element a of the group A which is mapped by f on the element b:

$$b = f(a)$$

We will determine the inverse image of the element b in the mapping f, i.e. the set of all elements x of the group A that are mapped by f on b. We usually denote this inverse image by $f^{-1}(b)$.

Thus $f^{-1}(b)$ is by definition the set of all elements x of the group A satisfying the equation

$$f(x) = b$$

As already stated, let a be an arbitrary element which is mapped on b. If x is another element of the set $f^{-1}(b)$ then we have

$$f(a) = b \quad f(x) = b \quad f(-a) = -b$$

$$f[x + (-a)] = b + (-b) = 0$$

(the zero on the right is the null element of the group B), and this means that $x + (-a)$ is a certain element u of the group U; thus $x = a + u$ is an element of that coset of the invariant subgroup U to which a belongs. Conversely if a and x lie in one coset then we have

$$x = a + u$$

$$f(x) = f(a) + f(u) = f(a) + 0 = f(a)$$

i.e. a and x are mapped on the same element b of the group B, or in other words they are contained in the same inverse image $f^{-1}(b)$.

Thus the inverse images $f^{-1}(b)$ of the elements of the group B are the cosets of the group A relative to the invariant subgroup U.

In this way there is set up a one-to-one correspondence ψ between the group B and the group V.

To each element of the group V, which is a certain coset of the group A relative to the invariant subgroup U, and therefore the inverse image of a certain element b of the group B, there corresponds precisely this element b of the group B. And each element b of the group B is associated with exactly one coset, i.e. with exactly one element of the group V, namely that coset which is the inverse image of b. The mapping ψ is homomorphic. Indeed let v_1 and v_2 be two elements of the group V and let

$$v_1 + v_2 = v_3 \tag{1}$$

Let a_1 be an arbitrary element of the coset v_1, a_2 an arbitrary element of the coset v_2, and $a_3 = a_1 + a_2$.

We know then that a_3 belongs to v_3.

We write

$$f(a_1) = b_1 \quad f(a_2) = b_2 \quad f(a_3) = b_3$$

Since f is a homomorphism, we have

$$b_1 + b_2 = b_3 \tag{2}$$

But since v_1, v_2, v_3 are the corresponding inverse images of the elements b_1, b_2, b_3 we have

$$\psi(v_1) = b_1 \quad \psi(v_2) = b_2 \quad \psi(v_3) = b_3$$

so that the equation (2) may be written in the following form:

$$\psi(v_1) + \psi(v_2) = \psi(v_3)$$

Thus we have proved that the mapping ψ is homomorphic. On account of the one-to-one nature of the homomorphic mapping of the group V onto the group B this mapping is an *isomorphism* of V onto B.

The final result of all these investigations is the following theorem:

The Homomorphism Theorem.—Every homomorphic mapping of a group A onto another group B has as kernel a certain invariant subgroup of the group A. Conversely every invariant subgroup U of the group A is the kernel of a certain homomorphic mapping φ of the group A onto the difference module V of the group A relative to U. We obtain the mapping φ if we associate with every element of the group A its coset with respect to the invariant subgroup U. If f is an arbitrary homomorphic mapping of the group A onto the group B, then the inverse images of the elements of the group B in this mapping are the cosets of the group A relative to the kernel U of the mapping f, and the group B is isomorphic to the difference module of the group A relative to U.

Therefore the invariant subgroups of a given group coincide with the kernels of all the possible homomorphic mappings of this group. All the groups which are homomorphic to A coincide with those groups which are isomorphic to the difference modules of the group A relative to all possible invariant subgroups of A.*

* It is left to the reader to reconsider, in the light of the homomorphism theorem which we have just proved, the examples of invariant subgroups and homomorphic mappings which were treated earlier, and to determine the difference modules corresponding to them.

Corollary.—A homomorphic mapping of a group A onto a group B is an isomorphism if and only if the kernel of this mapping consists only of the null element of A.

EXERCISES ON CHAPTER VIII

1. Verify that the permutations $\begin{pmatrix}1 & 2 & 3\\ 1 & 2 & 3\end{pmatrix}$, $\begin{pmatrix}1 & 2 & 3\\ 2 & 3 & 1\end{pmatrix}$, $\begin{pmatrix}1 & 2 & 3\\ 3 & 1 & 2\end{pmatrix}$ form an invariant subgroup H of S_3. Find the cosets of H and set up an addition table for the difference module $S_3 - H$. Describe the homomorphic mapping of S_3 onto $S_3 - H$ with kernel H.

2. G_{12} denotes the cyclic group of order 12 generated by a, and H is the subgroup generated by $3a$. Find the cosets of H in G_{12}, set up an addition table for the difference module $G_{12} - H$, and describe the homomorphic mapping of G_{12} onto $G_{12} - H$ with kernel H.

3. Find the elements of the difference module $G - H$ and write down its addition table, where G is the group of all whole numbers with ordinary addition as the group operation and H is the subgroup of all even numbers.

4. Prove that the factor group G/H, where G is the group of all non-zero complex numbers with ordinary multiplication as the group operation and H is the subgroup of all positive real numbers, is isomorphic to the group κ (see p. 46).

5. If G is the group of all complex numbers with ordinary addition as the group operation and if H is the subgroup of all real numbers, prove that $G - H$ is isomorphic to H.

6. Prove that the difference module $D_8 - C$, where D_8 denotes the dihedral group of order 8 and C its centre, is isomorphic to Klein's four-group.

7. Prove that a group of prime order is necessarily cyclic.

8. Prove that a subgroup of index 2 is necessarily invariant.

Appendix

ELEMENTARY CONCEPTS FROM THE THEORY OF SETS

The most important concepts of set theory, which we discuss in this appendix and which are being applied continually in mathematics, are in the first instance the concept of a *set*, of a *mapping*, and of a *partition*, as well as the elementary set-operations of forming the union and the intersection of several (sometimes of infinitely many) sets.

§ 1. The concept of a set

The concepts of a set and of a mapping belong to those mathematical concepts which cannot be described in terms of simpler concepts, and hence cannot logically be defined. Therefore we speak only of *explaining* the meaning of these concepts.

In everyday life as well as in every scientific study we are continually making use of the concept of a set, or, as it is often called, of an aggregate. We can speak of a set or aggregate of objects which are in a given room at a given time, of the set or aggregate of people who are present in the lecture-room or concert hall, of the set or aggregate of trees growing in a certain garden, of the set of books belonging to a given library, of the set of stars in the Milky Way, and so on. Further, we can speak of the set of molecules which are contained in a volume of given material, or of the set of cells in a living organism.

When we speak of a flock of geese, a sack of potatoes, a basket of apples, then from the mathematical point of view these are just sets: of geese forming the given flock, of potatoes or apples in the sack or basket.

The examples which we have given are examples of finite sets; that is to say, each of them is a set consisting of a certain *finite* number of elements, which may be a very large number (as for example in the case of water-molecules contained in a given volume of water), but which is always finite.

But infinite sets also occur. Such are for example the set of all natural numbers (i.e. positive integers), the set of all lines (in a plane or in space) which pass through a given point; the set of all circles through two given points, the set of all planes through a given line, and so on.

Set theory is principally concerned with the investigation of infinite sets.

The theory of finite sets is sometimes also called combinatorial analysis.

The simplest properties of sets, which we shall be speaking about here, almost always apply equally well to both finite and infinite sets.

We note next that in mathematics it is quite justifiable to consider sets containing just one element, as well as the set containing no element at all (the so-called " empty " set).

Let us suppose in the first instance that we are considering a set of circles passing through certain given points. If the number of these points is two, then the set of circles passing through them is infinite. However, if the number of points is three, then, provided the three points do not lie on a straight line, there is only one circle passing through them. In other words, the set of circles passing through three points consists of just one element. But the set of circles through three collinear points contains no element. It is the empty set since no such circle exists.

We explain this idea further by an everyday example. Suppose that we are speaking of the set of schoolboys who are present at a certain lesson and who are between 17 and 19 years of age. This set is completely determined in the sense that we can find out from each of the schoolboys present at this lesson, by making a simple inquiry, whether he belongs to this set or not. But evidently we do not know beforehand how many schoolboys do belong to the set. It may be ten, it may be five, it may be one, and it may be that there are no schoolboys of this age group in our class—if, say, they are all younger than 17. In this case our set is empty; in other cases it contains ten or five elements or one element.

Sets consisting of a single element will often appear in this book. Here it is not necessary for us to consider further the empty set; but it is often necessary and expedient to make use of it in mathematics.

§ 2. Subsets

We consider the set A of all people who are present in a certain lecture-room. Then the set of women present and the set of men present in the lecture-room provide examples of subsets of the set A.

Examples of other subsets of the set A are: the subset of those people who are not yet 20 years old; the subset of those people who are not yet 30 years old; the subset consisting of all those people whose heights are between 5 ft. and 6 ft.; the subset of all those people taller than 5 ft. 9 in.; the subset of all those people who live in London; the subset of all those people belonging to a certain profession or to a particular social class.

It is evident without more ado that certain of these subsets can consist of a single element; other subsets may happen to contain no element at all. But it can also happen that any one of the given subsets coincides with the whole set A, as for example, if all the people present in the lecture-room are women or if they are all not yet 30 years old. Moreover it can happen that certain of these subsets coincide with each other (if, for example, all the people in the lecture-room are women and all are younger than 30 years old).

The following is the general definition of a subset:

A set B is called a subset of a set A if every element of B is at the same time an element of A.

A subset of the set A is called *improper* if it coincides with the set A (in other words: the set A is regarded as one of its own subsets, which is called improper). If B is a subset of A, then we also say that *B is contained in A*, or that *A contains B*, and we write: $B \subseteq A$ or $A \supseteq B$. The sign $\subseteq$ is called the *inclusion sign.* The empty set is a subset of *every* set (also called improper).

We give further examples.

The set of all even numbers is a subset of the set of all whole numbers. The set of all whole numbers is a subset of the set of all rational numbers.

§ 3. Set operations

1. The union of sets

We now turn back to the example which we considered at the beginning of the previous paragraph.

From among all the people who are present in a given lecture-room, we consider the set M of all those people who satisfy *at least one of the following conditions*:

1. They are younger than 20 years old.
2. They are taller than 5 ft. 9 in.

In other words: to our set M belong all those people who are younger than 20 years old (regardless of their heights) and also all those people who are taller than 5 ft. 9 in. (whatever their ages). The set M is called the *union* of the following two sets: The set M_1 of all people present who are younger than 20 years old, and the set M_2 of all people present who are taller than 5 ft. 9 in.

The general definition of the union of two sets A and B reads: *The set consisting of all elements of the set A and of all elements of the set B is called the union of the sets A and B.*

Remark.—From the example given above we recognize that we can still form the union of sets when they have elements in common. Naturally it can happen that the sets M_1 and M_2 have elements in common, that is, that in our lecture-room there are people present who are younger than 20 and at the same time taller than 5 ft. 9 in.

In particular we remark that: *If the set B is a subset of A, then the union of the sets B and A coincides with the set A.* For example, if the set A consists of all people present in the lecture-room and not yet 30 years old and the set B consists of all those present who are younger than 20, then evidently the union of A and B coincides with A.

In a completely analogous manner we define the union of three sets, and of four sets, and so on. We can also define the union of infinitely many sets. All this is summarized in the following definition:

Suppose there is given an arbitrary finite or infinite class of sets. The set of all elements lying in at least one of the sets belonging to this class is called the union of the given class of sets.

By way of example let A_k be the set of all regular k-gons in the plane (with $k = 3, 4, 5, \ldots$), then A_3 is the set of all equilateral triangles, A_4 is the set of all squares, and so on.

The set of all regular polygons is the union of the sets $A_3, A_4, A_5, \ldots, A_k, \ldots$

We denote by B_k ($k = 3, 4, 5, \ldots$) the set of all regular polygons whose number of sides does not exceed k. Then B_k is the union of the sets $B_3, B_4, \ldots, B_{k-1}, B_k$, and the set of all regular polygons is the union of the sets B_k, $k = 3, 4, 5, \ldots$

Furthermore, evidently $A_3 = B_3$ and

$$B_3 \subseteq B_4 \subseteq B_5 \subseteq \ldots \subseteq B_k \subseteq B_{k+1} \subseteq \ldots\ldots$$

Remark.—The union of sets is also often called their sum.

2. The intersection of sets

Let M_1 be the set of people present in a lecture-room who are younger than 20 years old, and let M_2 be the set of people present in the lecture-room and taller than 5 ft. 9 in.

We understand by the *intersection* of the sets M_1 and M_2 the set of elements which belong to both of the sets M_1 and M_2, and therefore in our example it consists of those people present who are younger than 20 and at the same time taller than 5 ft. 9 in. Naturally this set can be empty.

In general by the intersection of the sets belonging to a given (finite or infinite) class of sets we mean the set consisting of those elements belonging to all the sets of the given class.

We remark that if $B \subseteq A$ then the intersection of the sets A and B is simply the set B.

Remark.—The intersection of sets is also often called their product.

§ 4. Mappings or functions

Let us suppose that a certain number of people are going into the theatre. At the entrance to the theatre they hand over their coats, etc., and receive in exchange a number under which their belongings are looked after in the cloakroom.

What is it that interests us mathematically in this very familiar situation?

What interests us is the following fact:

To every member of the audience in the theatre there *corresponds* (or *is related*) a certain object, namely the number which this person has been given in the cloakroom.

If we associate, in any way whatever, with every element a of a certain set A a particular element b of a certain set B, then we say that *the set A is mapped into the set B*, or that there is given a *function* whose *argument* runs through the set A and whose *values* lie in the set B. In order to signify that the given element b is related to the element a we write $b = f(a)$ and say that b is the *image* of the element a in the given mapping f, or that b is the value of the function corresponding to the value a of the argument.

We shall now investigate the different cases that can arise.

It can happen that for a certain performance all the tickets are sold. Then also there will usually be no empty place in the cloakroom. Not only has every member of the audience got a number but also all the numbers are used up among the members of the audience. In general mathematical terms this case may be expressed in the following form:

To every element a of the set A is related an element $b = f(a)$ of the set B, *and also every element of the set B is related to at least one element of the set A*. (The words in italics should make clear the application to our example and in particular the fact that every number is disposed of.)

In this case we call f a mapping of the set A *onto* the set B.

Why do we stress that every element of the set B is associated with *at least one* element of the set A?

Because it can happen that to different elements of the set A there is related one and the same element of the set B. In our particular example this means that *several people have given up their coats to be kept under one and the same number.*

The most important kind of mapping is the mapping of one set *onto* another. We easily arrive at this situation if we start from the general case of a mapping of one set *into* another. Indeed, let us suppose given an *arbitrary* mapping f of the set A into the set B. The set of all the elements of B each of which is associated by the mapping f with at least one element of the set A is called *the image set of A in the mapping f*; we denote it by $f(A)$. It is evident that the mapping f is a mapping of the set A *onto* the set $f(A)$.

These remarks allow us to restrict ourselves in future to the consideration of mappings of one set *onto* another.

In the example about the visitors to the theatre, A is the set of people who attended a certain performance, and $f(A)$ is the set of all wardrobe numbers which are in use.

Definition.—Suppose there is given a mapping f of a set A onto a set B. Let b be an arbitrary element of the set B. The set of all elements of A to which the given element b corresponds in the mapping f is called the inverse image of the element b in the mapping f. We denote this set by $f^{-1}(b)$.

In our example b is an arbitrary number in the cloakroom of the theatre. The inverse image of an element b is the set of all visitors to the theatre whose coats have been hung up under this number b.

We consider now the case that *under each number is hung only one coat,* so that therefore the inverse image $f^{-1}(b)$ of each element b consists of only one element of the set A. In this case the mapping of the set A onto the set B is called *one-to-one*.

We give another example illustrating the concept of a one-to-one mapping.

We imagine a detachment of cavalry. A horse belongs to every rider, and a rider sits on every horse. Therefore there exists a one-to-one mapping of the set of all riders onto the set of all horses (of a given detachment) and also a one-to-one mapping of the set of all horses onto the set of all riders (we speak always of the riders and horses of a given detachment).

This example shows that a one-to-one mapping of a set A onto a set B automatically gives rise to a one-to-one mapping of the set B onto the set A: If every set $f^{-1}(b)$, where b is an arbitrary element of B, consists of only one element a, then we obtain the mapping f^{-1} of the set B onto the set A if to every element b of the set B we relate the element $a = f^{-1}(b)$ of the set A. *We speak of f^{-1} as the mapping inverse to f.*

Therefore a one-to-one mapping of a set A onto a set B leads to the following situation: We unite every element a of the set A with a certain uniquely determined element $f(a)$ to form a pair. Then it appears that every element b of the set B is paired off exactly once, and indeed with the element a of the set A uniquely determined by b. If to every element b of the set B we relate the element a of A which is paired off with it, then we obtain a one-to-one mapping f^{-1} of the set B onto the set A, which is inverse to the mapping f.

Therefore, in a one-to-one mapping of one set onto another, neither set is in a more privileged position since *each* of the two sets is mapped onto the other in a one-to-one fashion. In order to emphasize this equality of status of the two sets we often speak of *a one-to-one correspondence between two sets* and understand by this the two one-to-one mutually inverse mappings of each set onto the other.

§ 5. Partition of a set into subsets

1. Sets of sets (systems of sets)

We can consider sets which consist of various kinds of elements. In particular we can consider *sets of sets*, that is, sets whose elements are themselves sets. We have already come across them when we introduced the definition of the union and intersection of sets. There we were speaking of the union and the intersection of classes or sets (containing a finite or infinite number of sets), and therefore indeed of sets of sets. We add to the examples given there some more, which are drawn from everyday experience.

The set of all sports clubs in London provides an example of a set of sets (each sports club being composed of its members); the set of all scientific congresses in a given year or in a given country, the set of all trade union organizations, the set of all military units (divisions, regiments, batallions, companies, platoons, etc.) of a given army are likewise sets of sets. These examples show that sets, which are elements of a given set of sets, can sometimes intersect and sometimes have no common elements. Thus, for example, the set of trades union organizations in a certain country would provide an example of a set of mutually disjoint sets, under the assumption that a citizen of the particular country could not at the same time be a member of two different trade unions. On the other hand the set of all military units of any army is an example of a set of sets of which several elements are subsets of the other elements: every platoon is a subset of a certain company, a company is a subset of a division, and so on.

The set of sports clubs in a certain town in general consists of intersecting sets, since one and the same person can be active in several clubs (for example in a swimming club and in a football team or in a skiing club).

Remark.—For ease of expression, instead of speaking of a " set of sets " we may sometimes use such a phrase as " system of sets " or " set-aggregate ".

2. Partitions

We obtain a very important class of set-aggregates if we consider *all possible divisions of an arbitrary set into mutually disjoint sets.* In other words, we suppose that a set M is given which is exhibited as a union of mutually disjoint subsets (of which there may be a finite or infinite number). These subsets are terms of the union and also elements of the given division of M.

Example I.—Let M be the set of all pupils of any particular school. The school is divided into classes which evidently form mutually disjoint subsets whose union is the whole set M.

Example II.—Let M be the set of all pupils who attend secondary schools in London. The set M can be split up for example in the following two ways into mutually disjoint subsets:

1. We regard the pupils of one and the same school as forming one term of the union* (that is we divide up the set of pupils according to schools).
2. We regard the pupils of one and the same year (in different schools) as forming one term of the union.

* Under the assumption that each pupil attends only one school.

Example III.—Let M be the set of all points of the plane. We choose an arbitrary straight line g in this plane and divide up the whole plane into lines parallel to g.

Remark I.—Those readers who know what a coordinate system is may suppose the line g to be one of the coordinate axes of this coordinate system (for definiteness we may say the x-axis).

Remark II.—If a given set M is divided into mutually disjoint subsets whose union is M itself, then we simply speak of a partition of the set M (or sometimes of a partition into classes).

Theorem I.—Suppose a mapping f of a set A onto a set B is given. The inverse images $f^{-1}(b)$ of all possible points b of the set B form a partition of the set A. The set consisting of the classes $f^{-1}(b)$ is in one-to-one correspondence with the set B.

The truth of this theorem is immediately evident: To each element a of a set A there corresponds by the mapping f just one element $b = f(a)$ of the set B, so that a belongs to the inverse image $f^{-1}(b)$. But this means firstly that the union of the inverse images of the points b is the whole set A and secondly that they are mutually disjoint.

The set of classes $f^{-1}(b)$ is in one-to-one correspondence with the set B, since to each element b of B corresponds the class $f^{-1}(b)$, and to each class $f^{-1}(b)$ corresponds the element b of B.

Theorem II.—Suppose a partition of a set A into classes is given. This partition gives rise to a mapping of the set A onto a certain set B, namely onto the set of all classes of the given partition. We obtain this mapping by associating with each element of the set A the class to which it belongs.

The proof of this theorem is already contained in its statement.

Example.—In considering the partition of the London schoolchildren (in Example II, 1) the mapping of the set A of all pupils onto the set B of all schools has already been indicated. Corresponding to each pupil is the school to which he belongs.

In spite of the self-evidence of the results stated in our two theorems, they did not at once find a place in mathematics in appropriate mathematical terminology. But as soon as this was found it assumed very great importance in the logical foundations of different mathematical disciplines, particularly in algebra.

3. Equivalence relations

Suppose there is given a partition of a set M. We introduce the following definition: We call two elements of the set M equivalent with respect to the given partition of M if they belong to one and the same class.

If we divide up the London schoolchildren according to schools then two pupils are " equivalent " if they attend the same school (although they may belong to different years). If we divide them up according to years then two pupils are " equivalent " if they belong to one and the same year (though they may attend different schools).

Our equivalence relation defined above possesses the following properties:

It is *symmetrical*: If a and b are equivalent, then so are b and a.

It is *transitive*: If the elements a and b are equivalent as well as the elements b and c, then also a and c are equivalent. (Two elements a and c, which are equivalent to a third element b, are also equivalent to each other.)

Finally we conclude that every element is equivalent to itself; this is the *reflexive* property of the equivalence relation.

Hence every partition of a given set defines, among the elements of this set, an equivalence relation which is reflexive, symmetrical, and transitive.

We now assume there to be a criterion, whose nature need not be specified, which allows us to speak of certain pairs of elements of the set M as *equivalent* pairs. All that we assume about this equivalence is that it possesses the reflexive, symmetrical, and transitive properties.

We prove that this equivalence relation defines a partition of the set M.

We denote by K_a the class of a given element a of the set M and understand by this the set of all elements which are equivalent to a.

Since our equivalence relation is by hypothesis reflexive, every element a is contained in its own class.

We prove that: *If two classes intersect* (that is, have at least one element in common) *then they coincide with each other* (that is, every element of the one class is at the same time an element of the other).

Let the classes K_a and K_b have the element c in common. Denoting the equivalence of any two elements x, y by $x \sim y$ we have, according to the definition of the classes,

$$a \sim c, \qquad b \sim c$$

whence on account of the symmetry $c \sim b$ and on account of the transitivity

$$a \sim b \tag{1}$$

Let y be an arbitrary element of the class K_b. Then

$$b \sim y$$

On account of the transitivity, and using (1), we have

$$a \sim y$$

so that y is an element of the class K_a.

Now let x be an arbitrary element of the class K_a. Then

$$a \sim x$$

and on account of the symmetry

$$x \sim a$$

therefore by the transitivity, and using (1),

$$x \sim b$$

Hence on account of the symmetry

$$b \sim x$$

which means that x belongs to the class K_b.

Therefore two classes K_a and K_b which have an element c in common evidently coincide with each other.

We have proved that the different classes K_a form a system of mutually disjoint subsets of the set M. Furthermore the union of these classes is the whole set M since every element of M belongs to its own class.

We repeat the results proved in this section and combine them in the following theorem:

Theorem III.—Every partition of a set M defines, among the elements of the set M, a certain equivalence relation which possesses the properties of being reflexive, symmetrical, and transitive. Conversely, every equivalence relation existing between the elements of the set M, and possessing the reflexive, symmetrical, and transitive properties, defines a division of the set M into mutually disjoint classes of equivalent elements.

EXERCISES ON THE APPENDIX

1. Let A, B, C, denote the interiors of three circles, each of which intersects the other two. Indicate by shading the following: $A \cup B$, $A \cap B$, $A \cup B \cup C$, $A \cap B \cap C$, $(A \cup B) \cap C$, $(A \cap B) \cup C$; where $A \cup B$ indicates the union of A and B, and $A \cap B$ indicates their intersection.

Verify that $(A \cup B) \cap C = (A \cap C) \cup (B \cap C)$.

2. Determine the union and intersection (*a*) of all circles with centre O, and (*b*) of all circles passing through the two points P, Q.

3. Determine the union and intersection of all circles of unit radius which lie entirely inside a given circle of diameter three units.

4. A_k denotes the set of all positive integral multiples of the whole number k. Determine the union and intersection of the sets $A_1, A_2, \ldots, A_k, \ldots$

5. Prove that the union of all equilateral triangles which can be inscribed in the circle centre O and radius one unit is the set of all points lying inside the circle, and that the intersection of these triangles is the set of all points lying inside the circle centre O and radius $\frac{1}{2}$.

State also the solution of the analogous problem for inscribed squares and other regular polygons, as well as for polygons circumscribing the circle.

6. In a certain city $\frac{7}{8}$ of the homes have a wireless, $\frac{1}{2}$ have television and $\frac{2}{3}$ have a vacuum cleaner. What is the least proportion of homes which must have all three? What is the greatest proportion of homes that can have all three?

7. Let the relation $\sim$ among the real numbers be defined as:

(*a*) $a \sim b$ if and only if $|a| = |b|$.

(*b*) $a \sim b$ if and only if $|a| \geqslant |b|$.

(*c*) $a \sim b$ if and only if $|a - b| > 0$.

(*d*) $a \sim b$ if and only if $1 + ab > 0$.

Which of these is an equivalence relation?

REFERENCES

General Algebra.

1. Albert, A. A., *College Algebra* (McGraw-Hill, First Edition, Second Impression, 1946).

2. Birkhoff, G., and MacLane, S., *A Survey of Modern Algebra* (Macmillan, Second Edition, 1953).

3. Weiss, M. J., *Higher Algebra* (Chapman & Hall, 1949).

Group Theory.

1. Kurosh, A. G., *The Theory of Groups*, English translation by K. Hirsch, two volumes (Chelsea, 1955).

2. Ledermann, W., *Introduction to the Theory of Finite Groups* (Oliver & Boyd, Second Edition, 1953).

INDEX